MOUNTAINS
OF THE HEART

A Natural History of the Appalachians

CHICAGO
REVIEW
PRESS

Scott Weidensaul
20th Anniversary Edition

For Kay Weidensaul

The a uthor w ould l ike t o t hank O xford U niversity P ress f or p ermission t o r eprint e xcerpts from "On a Monument to the Pigeon," in A Sand County Almanac by Aldo Leopold. Copyright © 1949 Aldo Leopold. Used by permission of Oxford University Press. The excerpt from The Graenlendinga Saga in Th e V nland Sagas, t anslated ly M gnus M gnusson a d Hermann Pálsson, copyright © 1965 by Magnus Magnusson and Hermann Piilsson, is reprinted by permission of Penguin Books.

Library of Congress Cataloging-in-Publication Data

Names: Weidensaul, Scott.
Title: Mountains of the heart : a natural history of the Appalachians.
Description: 20th anniversary edition. | Golden, CO : Fulcrum Publishing,
    2016. | Includes bibliographical references and index.
Identifiers: LCCN 2015039448 | ISBN 9781938486883
Subjects: LCSH: Natural history--Appalachian Mountains. | Appalachian
    Mountains. | Nature--Effect of human beings on--Appalachian Mountains.
Classification: LCC QH104.5.A6 W45 2016 | DDC 508.74--dc23
LC record available at http://lccn.loc.gov/2015039448

Printed in the United States

Published by Chicago Review Press Incorporated
814 North Franklin Street
Chicago, Illinois 60610
ISBN 978-1-93848-688-3

"*Mountains of the Heart* is a masterpiece encompassing Scott Weidensaul's mastery of language and nature. His ability to notice, catch, and distill fragments of beauty missed by others, makes this the sort of book yearned for by all who cherish wild things and wild places."

— *Audubon* Magazine

"...a wonderfully mature and ecological history of the Appalachian Mountain range."

— *Library Journal*

"Scott Weidensaul doesn't just write books, he lives them... more than good reads, they're journeys through time and space."

— *Pittsburgh Post Gazette*

Chosen as a Top Ten book by *Spin Magazine*

"A fine, literate, ramble along the ridges and valleys of the great spine of the Appalachians."

— John H. Mitchell,
author of *Ceremonial Time* and
Living at the End of Time

"Like John Muir, Scott Weidensaul naturally melds science and poetry."

— Marilou Awiakta,
author of *Selu: Seeking
the Corn-Mother's Wisdom*

Acknowledgments

I have been fortunate over the years to have known and learned from many of the best naturalists and biologists in the Appalachians, and much of the inspiration for (and information in) this book is a result of their friendship and help. The list is long, but I am especially grateful to Jim Brett and Laurie Goodrich at Hawk Mountain Sanctuary; Dr. Gary Alt, Arnold Hayden and Jerry Hassinger of the Pennsylvania Game Commission; Jeff Lepore; Rick Imes; Dr. Jim Bednarz; Kevin Ballard; and Dr. Scott Shalaway; several of whom also reviewed all or parts of the manuscript for accuracy. The errors that remain are mine.

For help in researching Belle Isle, thanks to Bill Carpenter, captain of the *Trudine Norman* in St. Anthony, and Karen Spencer at the Newfoundland Department of Tourism. A particular thank-you to Randy and Emily Campbell and Carl and Maude Elliott at the southwest Belle Isle light station, for their extraordinarily gracious hospitality.

For friendship beyond price, thanks to Margie Peterson, Rick Walters, Bruce Van Patter and Jeanne Tinsman.

A single tuliptree leaf rests in the branches of a tamarack.

Contents

Morning fog lifts in Cades Cove, in Great Smoky Mountains National Park.

Preface to the Twentieth Anniversary Edition

To a human, twenty years is a long time. It represents the span from infancy to adulthood—and for some, the shift from robust middle age to a retirement home (or even to the grave).

Not so with mountains. Time wears slowly on them, especially an ancient range like the Appalachians, whose history stretches back nearly 500 million years. For a system whose roots reach to before the dawn of modern fishes, and which rose skyward when the ancestors of dinosaurs and mammals were still diverging, twenty years is a comically short period. Compared with 500 million years, it barely counts as an eyeblink—a mere 0.000004 percent. If the history of the Appalachians encompassed a year, those twenty years would represent the final 1.26 seconds.

Yet we are fleeting, ephemeral humans, and we mark our anniversaries. It's been twenty years since *Mountains of the Heart* was first published, and if that is a tiny tick on the geologic clock, it's not without its significance. The two decades since this book was first published have brought many changes to these venerable hills, all of them wrought, to one degree or another, by the busy, bustling people that now inhabit every corner of the Appalachian landscape—and by many who live far away from it as well.

The publication of a fresh edition gives an author a chance to revisit a subject, if he chooses. There is a tension between preserving the flavor of the original work, which is a product of a particular time, place and mindset, but which may now be outdated or simply incorrect. That's an especially tricky balancing act for a book such as this that depends so heavily on first-person narrative. Consequently, I have tried to bring *Mountains of the Heart* into the twenty-first century with a light hand. Most remains as it was written in 1994, partly because so much of it deals with timeless subjects untouched by the passage of a few decades.

But the mountains have not themselves been untouched by the passage of twenty years. To hills that suffered so much in the past, from mining and timbering, abuse and pollution, there are new threats. Extractive industries have long had their way with these ridges and valleys (and with the people who inhabit them), but never—to choose one of the most egregious examples—on such a landscape-altering scale as the mining technique known as mountaintop removal.

It's difficult for anyone who hasn't seen a mountaintop removal operation to really appreciate the scale of the destruction—and because most of it occurs in poor regions far off the tourist track, most Americans never do witness it. The name sums up the devastation: With explosives and immense equipment, the upper third or more of a mountain—sometimes as much as eight hundred feet of ridge—is blasted to rubble. This overburden, which hides the thin seam of coal, is pushed into the neighboring valley, engulfing forests, entombing streams and burying homesteads that have been there for centuries. The streams now buried beneath the overburden seep out, full of toxins like selenium that leach from the smashed debris, ruining those waterways downstream that escaped direct destruction.

The result is a vast, flat wasteland of a plateau, sometimes up to fifteen square miles in size, bearing little but rough grass, where a rich Appalachian forest (and tight-knit Appalachian communities) once stood. By 2012, the U.S. Environmental Protection Agency estimated some 2,200 square miles—an area roughly the size of Delaware—had been destroyed by mountaintop removal, including two thousand miles of streams. The litany of environmental and health ills stemming from mountaintop removal is long and depressing—and extraordinarily well documented, which makes the decades-long struggle against mountaintop removal especially frustrating, despite some belated federal restrictions.

More recently, a wholly new energy boom has emerged in parts of the Appalachians in the form of shale gas extraction, especially in the Marcellus Formation underlying New York, Pennsylvania and West Virginia. New drilling techniques are providing access to gas trapped up to eight thousand feet below ground, in particular the approach known as hydraulic fracturing, or "fracking." In fracking, water, sand, salts and a witches' brew of hundreds of chemicals are pumped down under extreme pressure, fracturing the rock strata and freeing tiny bubbles of natural gas—revolutionizing gas extraction and bringing industrial energy production to regions that were just a few years ago quiet rural counties. Fracking has sparked enormous conflict, pitting neighbors against neighbors, local municipalities against states, and prompting concerns about the potentially devastating ecological problems associated with it.

Overlay a map of the Marcellus Formation with a map of Pennsylvania, and the correspondence between where the gas is most readily accessible, and the wildest, emptiest parts of the state—notably the Appalachian High Plateau country of the state's northern tier—is instantly apparent. That makes the impact of the drilling boom, and the reckless way in which it has often been managed, even more painful for a conservationist.

The drill pads, access roads and pipelines fragment what had been large, intact areas of forest; tanker trucks make hundreds of trips hauling water and equipment, compressor stations create ear-splitting noise and floodlights bathe drilling derricks in harsh light around the clock. Researchers have confirmed w hat l ocal r esidents h ave long claimed—that methane from gas wells has seriously contaminated groundwater in northeastern Pennsylvania, where drilling has been especially heavy. A 2011 study in the *Proceedings of the National Academy of Sciences* was for the first time able distinguish the chemical signature unique to deep-origin methane. The image of a hapless homeowner igniting tap water with a match has become an iconic feature of the new gas boom.

But the impact stretches far beyond poisoned water supplies. Each gas well requires up to 9 million gallons of water to frack, usually drawn from (and at times seriously depleting) surface sources like streams and rivers. Once used for fracking, millions of gallons of that water returns to the surface as "flowback"—water badly polluted with toxic drilling compounds (which energy companies, claiming proprietary rights, refuse to make public), frequently radioactive from radon contamination, and requiring specialized handling that few water treatment plants are equipped to provide. One method of disposal is to inject the waste into deep wells—causing, the U.S. Geological Survey estimates, a sixfold increase in earthquakes in areas where such wells are in use.

As bad as the shale gas boom could be for the central Appalachians, the ways in which some politicians have handled it have made the damage infinitely worse. Pennsylvania has become the poster child for the worst possible approach, treating its system of state forestland as a veritable cash machine to balance its budget by leasing more than seven hundred thousand acres for natural gas drilling. Even though the state concluded in 2010 that additional leases would harm especially sensitive areas within the state for-

Marsh marigold, false hellebore and skunk cabbage grow in the moist soil of Panther Brook, a stream in the New Jersey Appalachians.

ests, proposals for still further leases have been floated. (If there is an ironic silver lining, it is that the boom in inexpensive natural gas production has eviscerated the coal industry, and in the past few years, mountaintop removal has largely ground to a halt as some of the biggest coal producers have filed for bankruptcy.)

Yet this remains, as I argued two decades earlier, a "bruised but resilient land." In fact, the astonishing, regenerative powers of the natural world are on full display in the Appalachians—especially when, despite politics and inertia, we take steps to right the damage of the past. For instance, when *Mountains of the Heart* was first published, one of the most pressing and widespread environmental issues in the Appalachians was acid precipitation, which was severely damaging lakes and streams throughout the region—although there was hope, stemming from the changes in 1990 to the Clean Air Act, that the situation might improve.

And it has. Emissions of sulfur dioxide and nitrogen oxides from coal-fired power plants, the main source of acid rain, have declined substantially since the early 1990s, and with these declines have come significant reductions in acid precipitation. The improvement in aquatic health has been measurable in many areas, although the U.S. Geological Survey has warned that for some especially sensitive regions where natural buffering is limited, acid rain levels are still high enough to cause harm. But the trend on this once seemingly intractable problem is moving in a very positive direction.

Where wildlife is concerned, the news has been largely positive—with a few glaring and (in one case) frightening exceptions. Many once-rare species have become common, while others long missing from the mountains entirely have returned. And because these are complex organisms in complex environments, the results have not always been what we expected.

In 1999, I joined a group of scientists and wildlife managers at daybreak on a frigid, snow-swept ranch in the Wasatch Mountains of Utah. Overnight, the sweet scent of hay had lured sixteen elk into a huge, fenced corral trap, from which we were able to herd them into livestock trailers—the first step on a long road, literally and metaphorically, to restoring the species in the southern Appalachians.

That goal would have seemed a pipe dream twenty years ago, when I wrote about the small, beleaguered elk herd of the Pennsylvania Alleghenies—at the time, the only wild elk in the Appalachians. Elk have always been an especially iconic species to me, emblematic of the losses—and the tenuous links to the past—that these mountains embody. Nor am I alone. I've met many people to whom the notion of eastern elk is a powerful one. That pull rests, in part, in the seeming incongruity of a species so bound up in popular imagination with the West, but at home among the wooded hollows and long ridges of the East.

But in fact elk were once at least as common as deer in the Appalachians, ranging from southern New England to Georgia—until they were exterminated in the mid-nineteenth century, that is. When *Mountains of the Heart* came out, the only reminder of that lost monarchy was the tiny, long-neglected herd in that remote part of northern Pennsylvania, descended from Yellowstone elk released around World War I. Even that remnant came close to disappearing several times, at one point dwindling to just a few dozen animals.

Around the time *Mountains of the Heart* was published, though, a renaissance was dawning for elk in the East. The Pennsylvania herd, better protected and better managed, had grown to several hundred, and the state was tentatively exploring an expansion of their range by trapping and moving a few. Elsewhere, though, after decades of wistful speculation, elk restoration was becoming a much more forceful reality. A number of states in the East began feasibility studies in the late 1990s, and in 2001–2002, Great Smoky Mountains National Park reintroduced 52 elk to the Cataloochee Valley on the North Carolina side, where the herd has grown slowly to about 150 animals.

No one approached the matter of elk restoration with as much muscular action as Kentucky, however. In 1997, that state embarked on an audacious plan to move up to eighteen hundred western elk to the Cumberland Plateau, where biologists believed the combination of light human population, lots of publicly owned forestland and open areas on reclaimed strip mines (including old mountaintop removal sites) would create an ideal situation for elk.

My trip to Utah was part of that initial effort. As we stood outside the corral trap, our breath streaming in the zero-degree air, the biologists managing the work realized we had a problem. Among the cows and young bulls caught overnight was a "raghorn"—a smallish bull, five points on each side of his high antlers that would pose a danger to the others if he was penned up with them for the drive back to Kentucky.

Handlers were trying to herd the cows and youngsters into the cattle trailer while keeping the raghorn out, and at first it went well. About half the animals had been loaded when the bull broke away, dodging in a flurry of hooves and kicked snow, and ran straight down the loading chute.

"Cut him off!" someone yelled, and two of us standing by the chute grabbed the only tool at hand—a length of old burlap hanging from the end of the passageway—holding it like a matador's cape across the opening.

The elk jammed himself up against us, eyes wide and dark, steam pouring from his nostrils, his antlers clattering noisily on the metal sides of the chute. He smelled raw and wet with fear and anger, looking immense and very clearly dangerous. Yet against all logic, we held him at bay with that flimsy piece of fabric while the crew quickly closed the trailer bay and eased the vehicle away from the door, opening a gap through which we could at last let the bull escape, exploding away from us and up into the hills.

A week later, having cleared a battery of medical tests, the other elk we'd trapped were driven across the country in a single, thirty-one-hour trip. Along with a crowd of a couple hundred people, I was waiting for them at Pine Mountain, about twenty-five miles from the Virginia border. It was a gray, dark morning, windy and spitting rain, but the crowd was hushed and breathless as the first cow elk stepped carefully from the trailer. She hesitated, walked a few steps, then ran – and like water pouring from a jug, the rest of the herd raced out after her, forming a tight knot as they moved across the open grassland.

In all, the state moved about fifteen hundred elk—and its restoration dreams were exceeded in exceptionally short order. Eastern Kentucky is very nearly ideal for elk. The state's original goal of seventy-eight hun-

Gnarled trees, bent from the perennial wind, shade a flatrock outcropping near Grandfather Mountain, North Carolina.

dred animals was achieved more than a decade ahead of schedule, and the herd kept growing. Today it numbers about ten thousand, covering most of the Cumberland Plateau despite increasingly lenient hunting seasons, and has spread into neighboring regions of Virginia and West Virginia.

It has not been an unmitigated success, however—at least not in the eyes of many local residents. Elk hunting is now popular, and Knott County has billed itself as "the Elk Capital of the East" to take advantage of tourism, including an annual elk expo and horseback elk-watching tours. But elk are big, often inconvenient and occasionally dangerous animals. They raid gardens, have been known to rip the siding off homes while cleaning their antlers and attack livestock when in the hormone-soaked throes of the fall rut. They walk in front of vehicles, and as dangerous as hitting a hundred-pound deer with a car can be, a seven-hundred-pound elk is far worse.

"They're a danger to us, and we want them gone," one woman told the Associated Press after a bull elk she hit slid through the windshield of her compact car while she was on her way to Bible study. To mollify complaints, the state now allows landowners in some hard-hit areas to shoot any elk that show up on their property. Some of the problem elk have been shipped to Missouri, and across the border to Virginia—which, despite Kentucky's experience, is pushing ahead with its own restoration plan.

Elk are, however, only the most dramatic of many wildlife restoration stories in the Appalachians during the past two decades. Bald eagles, peregrine falcons and ospreys, all of which disappeared from most of the mountain system during the DDT era after World War II, have come roaring back thanks to pesticide bans and laborious reintroduction projects. Scientists tracking fishers with GPS-enabled collars discovered that the fox-sized weasels commonly prowl suburban backyards in Albany, while black bears now number in the tens of thousands in regions where a few decades ago they were almost a relic species. Wild turkeys are so common even in urbanized areas that homeowners worry about aggressive gobblers, hyped up on testosterone in the spring, roughing up little kids at the corner bus stop. Ravens, once the symbol of northern wilderness, have moved from the most remote mountaintops and empty cliffs to nest on football stadiums or the superstructures of electrical generating plants.

Not every trajectory is inexorably upward. When this book was first published, the future looked especially bright for moose, a species that had made a remarkable comeback in the Northeast. Populations in Maine, New Hampshire and Vermont were booming, and smaller numbers were found in the Adirondacks. By 2010, New York estimated that there were between five hundred and eight hundred moose in the state, some of them occasionally pushing south in the Catskills (from where they are usually trapped and moved back to the Adirondacks).

Moose are important not only as a reknitted ecological connection, but an economic engine. New Hampshire, whose moose population grew from just fifty in the 1950s to more than five thousand animals in recent years, estimates that moose-watching is a $115 million boost to tourism. In Maine, moose are as much a cultural and regional icon as are lobsters. But the good times for moose may be over. Not only in the Northeast but right across the southern rim of the moose's range—from Quebec to Minnesota, Montana and British Columbia—wildlife biologists have been documenting a dramatic collapse in moose numbers.

One reason may be an explosion in a parasite called the winter tick, which increasingly infests moose—about one hundred fifty thousand have been found on a single animal. Moose, which unlike deer don't instinctively groom off the pests when they first appear, are later so irritated that they scratch themselves almost naked, then perish from hypothermia, anemia and malnutrition. Such "ghost moose," named for the pale gray of their exposed underfur, are an increasingly common and grim sight in the northern Appalachians.

Scientists point out that the ticks are actually just symptoms of a deep problem. Steadily warmer winters, and especially mild, snow-free autumns, —the effects of climate change—have allowed this native parasite to reach epidemic proportions. Similar issues have arisen with other parasites like liver flukes that require milder conditions. Warmer winters also pose a more direct threat to moose, which are so immense and well insulated that winter temperatures above about 23 degrees Fahrenheit pose serious overheating problems.

The winter tick is a native species, but ours is an increasingly small, interconnected world in which pests and pathogens ricochet around the globe with frightening speed, and Appalachian flora and fauna have paid a heavy price for that. The tiny insect known as the hemlock woolly adelgid, which was just appearing in the Appalachians when *Mountains of the Heart* was published, has made good on its threat to the eastern hemlock; millions of these conifers, an ecological keystone in the Appalachians, have died. Still other pests—emerald ash borers and Asian long-horned beetles among the worst—have appeared as well.

The mountain range's globally significant diversity of salamanders has, thus far, escaped the worst of the chytrid fungus—a disease essentially unknown when the book was published, and which has now infected a third of the world's amphibian species, driving as many as a hundred species to extinction and hundreds more to the brink. No cure, treatment or prevention has been found, and the "death wave" of chytrid infections continues to roll through many parts of the world.

Chytrid has been called the worst infectious disease ever to hit a group of vertebrates, and in terms of the number of species and the geographic range it affects, that is true. (And the appearance of a salamander-specific chytrid in Europe, spread via the exotic pet trade, makes the danger to the Appalachians' salamanders even more grave.)

But even as biologists reeled from chytrid's impact on amphibians, a wholly new disease emerged among bats—and in this case, the Appalachians have been the epicenter of this new and horrifying epidemic.

In the winter of 2006–2007, a recreational caver noted something odd in Howe's Cave in Schoharie County, New York. Hibernating bats in the cave had a strange, white growth around their muzzles, as though they'd dipped their faces in confectioners' sugar. But it wasn't candy; it was *Pseudogymnoascus destructans*, more commonly called white-nose syndrome, a never-before-seen, cold-hardy fungus that has emerged as one of the worst—perhaps *the* worst—wildlife epidemic in history.

Bats infected with white-nose develop the characteristic fuzzy fungal growth on the muzzle, wings, tail or ears while huddled by the thousands to hibernate in densely packed masses. The disease seems to irritate the bats, rousing them from hibernation and burning up their carefully hoarded fat reserves; many bats, hungry and desperate, fly out of the caves into the winter cold, and are found dead by the entrance.

From Howe's Cave, white-nose syndrome began to spread with horrific speed while biologists scrambled to identify the cause and seek a cure. The fungus proved to be genetically almost identical to a European disease, to which European bats seem to enjoy immunity; the assumption is that humans brought *P. destructans* to this continent, perhaps on the boots and equipment of unsuspecting cavers. After the initial infection, it has spread naturally and rapidly by the bats themselves.

In just six years, white-nose syndrome killed almost 7 million bats in eastern North America, reducing even the most common species, like little brown bats, by as much as 98 percent. Hibernacula (the caves in which bats spend the winter) that once held tens of thousands of bats now held only hundreds, or dozens. In Pennsylvania, a mine that once held seventy-five thousand hibernating bats was down to fifteen hundred by 2013—and that was considered good. Another mine that just a few years ago hosted ninety thousand was down to ten—not ten thousand, but ten individual bats.

At this writing, white-nose syndrome has spread the length of the Appalachians from Alabama to New Brunswick, and west as far as Oklahoma and Minnesota. Biologists fear most for already rare species, like the federally endangered Indiana bat and gray bat, but the U.S. Fish and Wildlife Service has ruled that two formerly widespread species, the small-footed and northern long-eared bats, both also warrant protection under the federal Endangered Species Act. It's possible, perhaps even inevitable, that others will join them, including the little brown bat. That this, the most abundant bat in North America just a few years ago, could be brought to such straits so quickly is a lesson in the nightmarish potential that invasive pathogens hold.

You have to go back to the turn of the twentieth century to find an analogue to white-nose syndrome, in which an introduced disease upended the Appalachian ecosystem so quickly and profoundly. The chestnut blight from Asia hit with almost as much speed, and shrugged off all desperate efforts to contain or control it, destroying what was arguably the most economically and ecologically important tree in the East.

In 1994, hope for the American chestnut still seemed maddeningly distant. This cornerstone of Appalachian hardwoods was the focus of grueling restoration efforts lasting decades—especially the excruciatingly slow process of hybridizing American trees with smaller but blight-resistant Asian chestnuts, then painstakingly backcrossing them, generation after generation, to weed out all but the genes for blight resistance.

But twenty years on, that effort seems on the verge of success—as does a thoroughly modern approach involving gene splicing: lifting genetic material from wheat and inserting it into the chestnut's genome to confer blight resistance. Both efforts are in their final stages of testing, and the long-sought restoration of this majestic and much-loved tree may finally be at hand—and with it, a small sense of hope in the face of other swift and seemingly permanent losses.

In fact, we stand today at the cusp of a time when the very notion of permanent loss may be changing. Twenty years ago, I wrote about the many things I mourned in this later, perhaps lesser age. "I mourn the loss of the mighty forests, the vanished elk and bison that no longer tramp the Appalachians. I wish in vain for a wolf's song on a midwinter's night and must be content with a yapping fox."

Now, the elk are back, and a few pioneering wolves have even returned to the Canadian end of the range. Mountain lions may be slipping in from the West. But some things, logic tells us, are gone for good—and foremost among them, to my grieving Appalachian soul, is the passenger pigeon.

"I have sat with eyes closed and tried, straining my imagination, to re-build in my mind the multitudes that gathered the sky to themselves and blew a wind beneath their assembled wings like the roar of a storm," I wrote in *Mountains of the Heart.* "But it cannot be. Some things we lose but still remember; others pass away into a limbo that defies recall. The pigeon is in this farther recess, reduced to quiet words on a page, growing fainter with the years."

On October 8, 2013, scientists in California flipped a switch on an Illumina HiSeq 2500 gene sequencer and began sifting through the DNA of a female passenger pigeon that was shot and stuffed in 1871 near Toronto. When it was finished just forty-eight hours later, the machine had decoded all 1.3 billion base pairs of the dead pigeon's genome—the first step of an audacious and controversial plan to literally resurrect this extinct bird.

The term bandied about today is *de-extinction*, the idea that lost species can be brought back from the dead through molecular biology. To hear de-extinction proponents, it is a matter of when, and not if—a question only of when scientists figure out how to reengineer the genome of the passenger pigeon's nearest living relative, the band-tailed pigeon; of when they can create artificial chromosomes using that altered genome, form them into a nucleus, and insert that into a band-tailed's egg cell; a matter of when, and not if, from this synthetic beginning they re-create a living, breathing passenger pigeon. When, and not if, they mine the thousands of pigeon skins and mounts in museums to clone genetically distinct, genetically diverse flocks of de-extinct pigeons to once again flood the Appalachian skies.

Perhaps. Cloning birds is challenging enough that no one has yet done it, even with living species. But neither do I underestimate the speed with which molecular science has progressed. When I wrote a book about the search for lost species in 2002, which dealt in some depth with this kind of genetic tinkering, I suspected we were many decades away from that reality. Today I believe that successful resurrection is almost nigh—though the most immediate successes will be less dramatic, such as cloning recently extinct frogs from frozen tissue.

But we are, for the first time, confronted with the possibility of breaching what had been the ultimate form of death—extinction. That so much attention and effort has been focused on the passenger pigeon is ironic, since few extinct species would pose as many ethical and ecological challenges to restore as this, once the world's most abundant land bird. In a country with nearly 400 million people, is there room for a bird that numbers in the billions? Would humans tolerate the disruption, the potential for crop losses, even the basic mess of a bird whose droppings were said to have fallen thick and deep as snow? Is it even possible to create enough test-tube passenger pigeons to jump-start a wild flock, given that this bird's basic survival strategy entailed mega-sociality, and that anything less than millions might not be sustainable?

More fundamentally, are the Appalachians today suitable for this bird from the past? Is there sufficient habitat, or enough food to feed the multitudes? One of the scientists at the forefront of the passenger pigeon project has spoken eloquently about how the pigeons and the eastern hardwood forests once held each other in a great and elegant dance, and that the forests await their onetime partner. But the species to which the passenger pigeon were most tightly bound were masting trees like beeches and oaks, and especially the American chestnut, with its annual, predictable wealth of nuts. Perhaps, one expert on chestnut restoration has dryly observed, those trying to resurrect the passenger pigeon should wait seventy or eighty years for a new forest of chestnuts to rise.

I am not holding my breath—and I tend to agree with those more level-headed scientists who argue that, should science eventually make avian de-extinction possible, it makes more sense to start with two other longgone species of the northern Appalachian coast. Neither the great auk nor the Labrador duck, which vanished from the north Atlantic in the nineteenth century, pose the kind of thorny problems that the passenger pigeon raises.

The very fact that we can have this conversation, though, is an indication of the strange days in which we live. Twenty years is a heartbeat for an ancient mountain chain, but given what the past two decades have brought to the Appalachians, I can only imagine what the next two hold in store for these old and weathered hills—and for those of us who love them.

Scott Weidensaul
September 2015

Sheets of water, freshened by a night of rain, cascade down a cliff in Great Smoky Mountains National Park.

INTRODUCTION

The ridge is the first thing I see every morning and the last thing I see at night, a constant and comforting wall whenever I look across the fields and woodlots of the neighboring farms.

On a hazy summer afternoon it's a long, low swell that crowds the horizon, deep blue and only slightly more substantial than the humid air. On a crisp morning in early autumn when the wind is roaring in cold from the north, it is etched in amazing detail—the oaks with their leaves up-ended in the blast, shining silver against the green like frost, and the dark spires of the hemlocks and white pines marking where the small streams squirm down the mountain, hiding wild trout and grouse. In spring the oaks turn the furry color of buckskin, and on winter nights when the shadows lie black and the owls call, I can see the ridge only as the line where the stars vanish.

The Kittatinny Ridge is Appalachian, part of the same grand chain of mountains that nudges the clouds in other parts of the East. But it is no great peak itself. Here in central Pennsylvania the Appalachians sink low and gentle, like a swaybacked horse; the highest point is less than two thousand feet above sea level, barely a foothill when held against the White Mountains or the Smokies. Yet beneath the veneer of oak trees is rock that was formed when life was just learning to live on dry land, and which crumpled into mountains hundreds of millions of years ago. Such a legacy exerts a tremendous pull on a fleeting mortal like me.

The Appalachians captured me as a child, and they have never released their hold. I am not unusual in this; mountains in general seem to exert an especially profound grip on the human imagination. Whatever the reason, I get twitchy when I spend too much time in a place where the highest point on the horizon is a telephone pole or a grain silo. I need to be able to look into the hills and know that I could disappear into them when the tame world gets to be too much, like a promise of refuge always waiting on the doorstep.

I never spared much thought for the mountains when I was a child, except to use them as a wonderful playground. My friends and I scrambled over the sandstone boulders, fighting mock battles through the oaks and hickories, rolling over mossy logs to search for ringneck snakes or salamanders. The mountains were simply there, taken for granted—until I started watching the hawks that each autumn migrate south along the ridges of Pennsylvania.

The birds were a revelation of beauty and grace, but more than that, they opened my eyes to the true sweep of the Appalachians. These weren't just low, homey hills—they were part of a much greater system that flowed north and south from me, joining worlds I'd never even known existed. The magnitude of that discovery knocked the wind out of my twelve-year-old body.

I still find the Appalachians breathtaking. The chain stretches, nearly unbroken, across a course of more than two thousand miles, one of the greatest geographic features of this continent. At the southern end a naturalist finds longleaf pines, painted buntings and wild turkeys; at the northern end, woodland caribou, Arctic foxes and even (when the pack ice closes in on Newfoundland) a wandering polar bear or two. Few other continuous mountain chains in the world, and none in North America, encompass such biological diversity.

These hills are a sweep of time as well as distance. They are old, not just as humans reckon age, but as the Earth itself does—spawned by colliding continents caught in a global dance, buckled into peaks before the dinosaurs reached their ascendancy, ground down by hundreds of millions of years of wind and rain, polished by mile-thick glaciers.

On this vast and ancient framework, life has woven a fabric of astonishing intricacy. North and south, east and west meet in the Appalachians. On a single mountain, separated only by altitude, can be found rich southern cove forests with towering tuliptrees and Carolina silverbell; stands of northern hardwoods like sugar maple and yellow birch; and alpine conifer zones of spruce and fir, more akin to Canada than Dixie.

A mountain chain is the sum of many parts, and this is especially true of the Appalachians. They include five main geographic divisions: the Piedmont, the area of low, hilly country on the eastern flank between the mountains and the flat coastal plain; the Blue Ridge province, running from the Great Smokies up in a high, narrow slice through Virginia's Shenandoah National Park to South Mountain in Pennsylvania; the Ridge and Valley province, with its neatly parallel hills extending from southern New York to central Alabama; the Appalachian Plateau, the wide expanse of old, deeply eroded landscape that forms the western boundary of the mountain system; and the northern Appalachians, from the Hudson Valley north through Atlantic Canada and the western mountains of Newfoundland.

No one argues about where the Appalachians start in the South—their fringe of upland rises, like the train of a graceful skirt, from the farmland of central Alabama. From here the mountain spine flows north and east through northern Georgia and the western tip of South Carolina; along the Tennessee–North Carolina border; up western Virginia, eastern Kentucky and West Virginia; and across the heart of Pennsylvania. It skirts northern New Jersey, eastern New York and western Massachusetts, then rises through Vermont, New Hampshire and Maine. The Appalachians spread across the Canadian Atlantic Provinces like a river delta, a muddle of low hills but for the Shickshocks in Quebec's Gaspé Peninsula and the sudden cliffs of Cape Breton Island.

You can approach the question of where the Appalachians end in the north in two ways. Most maps (and many geographers) call a halt to the matter on the Gaspé, but geologists correctly point out that Newfoundland's mountains are Appalachian, formed by the same tectonic forces that created the rest of the chain. And so I followed the chain across the Gulf of St. Lawrence and up the Long Range Mountains on the great island's west, three hundred miles of ice-carved rock, where the stunted spruces are relegated to the lowest hills only, and most of the mountains stand bare before the wind.

The Appalachians finally give up their fight with the Labrador Sea about fifteen miles off the northern tip of Newfoundland, where a sliver of land known as Belle Isle breaches the waves like one of the many whales that hunt these frigid waters. It is here that the Appalachians finally end, here that the land dives deep and the ocean wins.

I've spent much of my adult life wandering up and down the Appalachians, trying to understand the mountains I love so deeply. With the miles and the time I've learned a little about the astounding web of geology, climate, evolution and 500 million years of history that produced the Appalachians, but I'm no closer to really knowing their secrets. Probably I never will, but that's all the more reason to keep trying.

At times in my travels I felt like a seeker after lost glories—yearning to glimpse the woodland sea that once lapped, almost unbroken, from one end of the mountains to the other, but having to be content with the scattered pieces of ancient forest that remain; wishing that I could see the multitudes of wild pigeons and the herds of native bison, or hear the howl of gray wolves in the high peaks. But they are gone, and will not come again.

The Appalachians are not pristine wilderness; they've harbored humans for thousands of years and have been radically altered by our presence. The pressures have mounted steadily—logging, mining, damming, development, now the insidious effects of pollution and disease that threaten the forests themselves—but so far the Appalachians have bounced back surprisingly well. Where the logging crews scalped the land a century ago the trees have regrown, and if the woods are not as towering and expansive as they once were, their very existence in the heavily populated, heavily industrialized East is almost a miracle. The healing continues; within my lifetime the bears have returned to the ridge that I see from my window, and so have the goshawks and the ravens and the bobcats. This is a bruised but resilient land, if we give it a chance.

A few words about what the Appalachians are not. They are not Appalachia—or at least not entirely. In fact, that term *Appalachia* will not appear elsewhere in this book, since the region to which it is usually applied makes up less than a quarter of the Appalachian chain. The Adirondacks are not Appalachian, despite their position snuggled in the corner between New York's plateau country and the Green Mountains. Jumbled and rough, they are a piece of the ancient foundation known as the Canadian Shield, rock more than a billion years old. The mountains themselves, however, were only uplifted within the last 10 or 15 million years—just youngsters compared to their Appalachian neighbors.

Nor does the famous Appalachian Trail play more than a passing role in the chapters to follow. The trail begins in Georgia, more than 150 miles from the true southern rim of the Appalachians, and ends on Mount Katahdin in central Maine, nearly 650 miles from the range's terminus in Newfoundland. Yet for many people, the trail so defines the Appalachians that they are surprised to hear that the mountains extend beyond the trailheads.

These are mountains of many names. The chain as a whole is referred to as the Appalachians, but there are dozens of smaller, discreet mountain systems contained within it. Some, like the Great Smokies in the south and the White or Green mountains in New England, are famous in their own right, but others are little known outside their own shadows. For me, their names are magically evocative: the Black Mountains, the Unakas, the Shickshocks, the Notre Dames, the Unicois, the Snowbirds, the Cold Hollow Hills. All ring with mysteries, and all hold secrets. They comprise a land with an uncanny ability to hold the human spirit—the mountains of the heart.

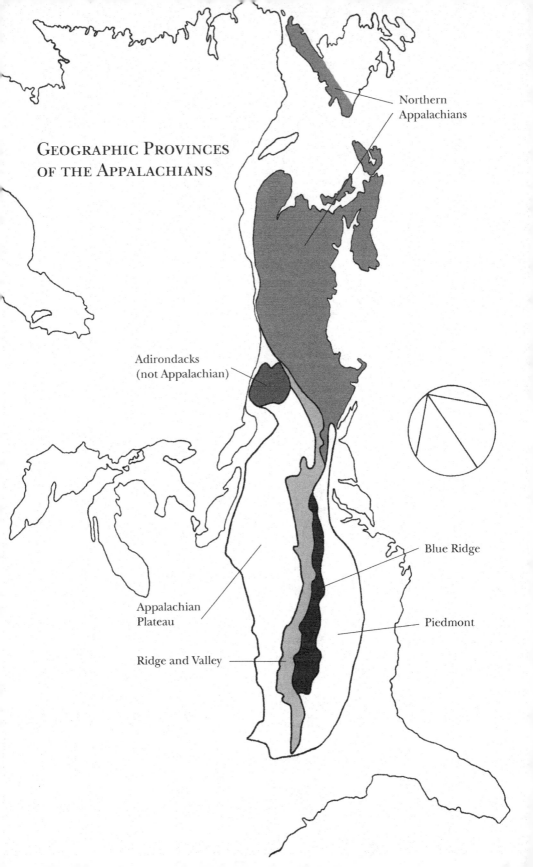

GEOGRAPHIC PROVINCES
OF THE APPALACHIANS

Northern
Appalachians

Adirondacks
(not Appalachian)

Blue Ridge

Appalachian
Plateau

Piedmont

Ridge and Valley

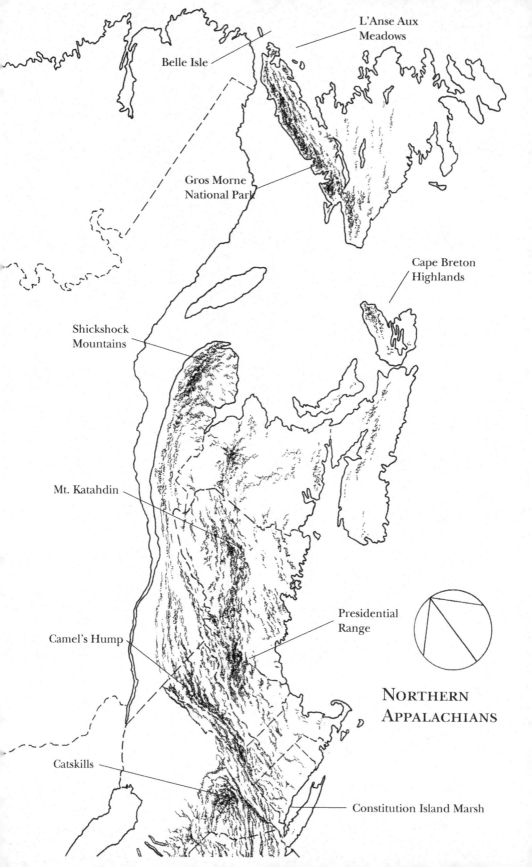

L'Anse Aux Meadows

Belle Isle

Gros Morne National Park

Cape Breton Highlands

Shickshock Mountains

Mt. Katahdin

Presidential Range

Camel's Hump

Catskills

Constitution Island Marsh

NORTHERN APPALACHIANS

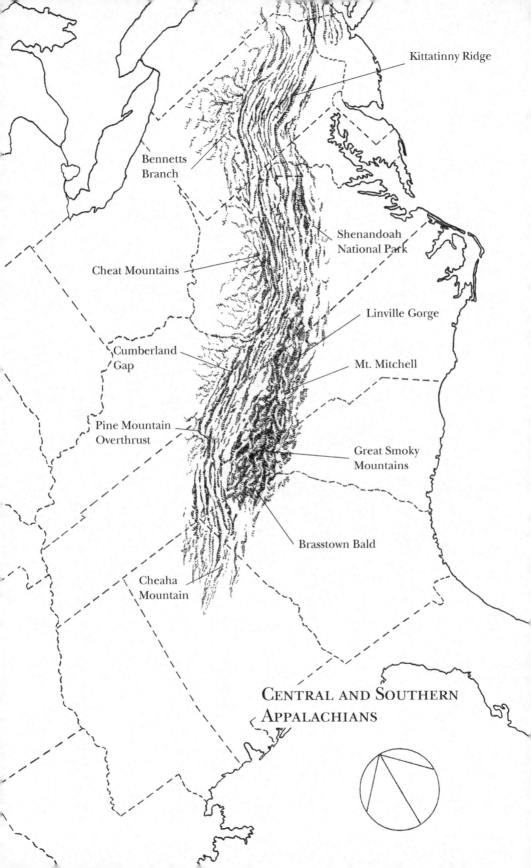

Kittatinny Ridge

Bennetts
Branch

Shenandoah
National Park

Cheat Mountains

Linville Gorge

Cumberland
Gap

Mt. Mitchell

Pine Mountain
Overthrust

Great Smoky
Mountains

Brasstown Bald

Cheaha
Mountain

**CENTRAL AND SOUTHERN
APPALACHIANS**

Birdsfoot Violets

Prologue

CHEAHA MOUNTAIN, ALABAMA

Blowing steady from the southwest, the breeze carries the smell of spring—the musky, green aroma of humus and moisture and growth. The scent drifts across the forest in the valley below, carried from the soybean and cotton fields of Alabama's fertile Black Belt a hundred miles away.

The breeze flows from the sunset, from a sky laden with straps of orange and high-flung horsetails of gray that foretell rain by daybreak. It flows across the tops of the oaks and the shortleaf pines that spread over the hills, which rise from the hazy distance to sharp-edged clarity far below my seat.

The wind tugs at my hair, but it cannot reach me in this narrow crevice of stone where I've wedged myself. The rock is split here into great, squared-off blocks that form cliffs along the edge of Cheaha Mountain, overhanging the valley at an alarming angle and providing a perch for pines so sculpted by wind and gravity that they are almost too perfectly gnarled.

Cheaha is the highest spot in a state not known for its mountains, a 2,407-foot hump in Alabama's northeast, about thirty miles from the Georgia border. It is not the southernmost point of the Appalachians, but on a clear day you can just about see that point from here, off to the west and south toward Birmingham, from whence blows the evening wind. The sunset light, filtering through the haze, has made the neighboring ridges translucent. The hills retain something of winter's brown, but the valleys have given themselves over to spring—canopies of kelly green and darker viridian where the hardwoods and conifers mix.

The windy solitude of this precipice comes as a relief; I've arrived on the same weekend as a major bike race, and the state park that encompasses the mountain is swarming with people. I had hoped for something wild and untrammeled and found instead all the worst kinds of park development—the concession stands with fake Indian trinkets, the mountaintop lodge and restaurant, the clusters of antennas bristling from the flat summit, the glass-enclosed observation tower. Even along the two main roads up Cheaha, the native birdsfoot violets, bluets and crested irises have been supplanted by thickly seeded strips of Italian clover, the sort of "wildflower" planting that passes for highway beautification.

But then, not long before sundown, I found this spot with its view of the Appalachians' southern anchor, and my spirits lifted. I've always found peace in settings like this, with my back to great slabs of rock and empty space at my feet: solidity and vapor. Tomorrow I start north again, the final lap on travels that have taken me up and down the spine of the Appalachians, from the flower-spangled cove forests of the Smokies to a tundra island off the coast of Newfoundland. But for now there is only the mountain, the feel of the chilly quartz through my shirt, the soft wind and the sweet light of sunset polishing the day. Time enough later for journeys; for now, the ancient pace of the mountains is enough.

A whitetail buck, his velvet-covered antlers just emerging, pauses in a meadow in the southern Appalachians.

Linville Gorge

Chapter 1

THE SUPPLE ROCK

I flicked a big, glossy black carpenter ant off my leg, shifted slightly against the pine tree at my back and thought about rocks.

The view from the lip of the Linville Gorge invites deep thinking. The Linville River, which spills out of western North Carolina's Blue Ridge, has cut a drastic slice into the earth here, a drop of more than a thousand feet to the sliver of water way down below, crowded by canyon walls wrapped in pine, oak and hemlock. It was a warm, dry, hazy day, and the woods smelled heavily of pine—an intoxicant after a long winter. The nose misses odors during the cold months but doesn't realize it until spring uncorks the perfume.

Rocks are fundamental to the Appalachians, but unless you are a geologist or a weekend rock hound, you're unlikely to spend much time thinking about them. For most of us rocks are just part of the scenery—something to climb, or stub our toes on, or skip across a quiet lake.

And when the nonexpert tries to concentrate on rocks, as I was doing, they often mock the effort. Of all the earthbound sciences, geology may be the most humbling. No other discipline save astronomy asks us to step so far outside of ourselves, to think in such desperately inhuman lengths of time or on so confoundingly huge a scale.

The Linville River has a fairly short shot at glory. It flows down through the gorge for about eighteen miles, then emerges into the gentle Piedmont hills, where it is immediately swallowed by the waters of Lake James, the westernmost impoundment on the Catawba River system.

But it makes the most of its brief freedom. Depending on how you measure such things, Linville Gorge is the deepest canyon east of the Mississippi, in places measuring more than seventeen hundred feet from riverbed to mountaintop. One would expect such a vertical cut to reveal the ordered layers of rock, much as the Grand Canyon's painted sediments trace billions of years of geologic history.

But in the Appalachians geology is a squirrely science, never giving you exactly what you expect. Take the Linville Gorge, for instance. At the head of the chasm, the river crashes down over a spectacular series of waterfalls, tier after tier of foam that booms in the enclosed valley. The rock over which it flows is metamorphic; that is to say, incalculable heat and pressure deep inside the Earth's mantle (where this rock once lay) converted the original granite into a new form, known as *gneiss*. This transformation took place about a billion years ago.

The water pours through a narrow slot in the high gneiss walls, makes a turn and fans out in a final cascade to a deep pool. The rock underlying the pool—and the waterfall—is actually *younger* by 500 million years than the gneiss above it.

In a well-ordered world where geology was easy, this sort of thing wouldn't happen. But tracing the geologic history of the Appalachians is, as some geologists have pointed out, like reconstructing a hit-and-run car wreck by examining the dents and the crumpled metal, peering at the twisted shards and looking for scrapes of paint from the other car. The southern Appalachians are a particularly nasty dent, and the culprit was the bulge of Africa, which rear-ended North America about 290 million years ago. The impact shoved much of the continental rim inland, piling it up in untidy heaps. Thus, older rock was pushed on top of younger rock, which—at Linville Falls and elsewhere—is sometimes exposed by erosion.

This incoherence of rock has driven geologists to distraction. On a map the Appalachians form a cohesive whole, strung out almost seamlessly over two thousand miles, but different regions of the chain have radically different ancestries. The age of the rock differs wildly; some of it is sedimentary, some igneous (like granite, made up of once-molten magma) and some metamorphic. The rock is far older than the mountains, just as the rock from which a stone house has been constructed is far older than the building. Nor were all the ranges of the Appalachians uplifted by the same forces or at the same time.

Geologists have long known that the Appalachians arose in distinct pulses spread over hundreds of millions of years, but the mechanism that pushed them skyward was unclear. Books written before about 1965 speak vaguely about accumulated heat and pressure, and about the tendency of lighter crustal zones to buoy up. "Many mountains and plateaus are

Hardy pines cling to the sandstone cliffs at Cheaha Mountain, Alabama, near the southern terminus of the Appalachians.

high because they are light," says one volume from the 1940s on the northern Appalachians, leaving the subject at that. The less said, apparently, the better.

That changed by 1968, when the theory of global plate tectonics came to prominence. At its simplest, tectonics theory suggests that the Earth's outer crust, or lithosphere, is made up of enormous, distinct plates (seven to ten major slabs, depending on who is counting, and a bunch of smaller platelets), which float on denser layers of molten and semi-molten rock. As these plates move, they heave and buckle, hitting each other to raise mountain ranges and tearing apart to form oceans. Heavier oceanic plates, made up of dense basaltic rock, tend to slide beneath the somewhat lighter continental plates of granite when the two meet. The process, known as subduction, is happening now along the Pacific Northwest coast, where the dense Juan de Fuca Plate is subducting beneath the granitic North American Plate, forming inland volcanoes like Mount St. Helens.

The idea that the continents were once joined together and have since moved to their current positions has been around for a long time—much of it as material for jokes. In the nineteenth century, the theory was ignored in favor of "the doctrine of permanence," which held that continents and oceans might change shape slightly as the Earth contracted, but would not move in any meaningful degree. When first proposed scientifically in 1912 by German meteorologist Alfred Wegener, the idea of drifting continents was ridiculed, even though Wegener supported it with data indicating that South America's eastern bulge and Africa's western bight hold similar rock formations and fossils. And, of course, the continental shapes of that bulge and bight fit together like pieces of a puzzle. Unfortunately, Wegener could not supply a convincing mechanism for all this continental shuffling, and his theory won few converts.

It took nearly fifty years for the idea of continental drift to reemerge, this time largely due to exploration of the sea floor. Enormous, mid-ocean rifts had been discovered, with the spreading out to either side at a rate of two inches a year as magma welled up and solidified, while the edges were being sucked beneath other plates. It was found that as molten rock cools, particles within it line up with the Earth's magnetic field, allowing geologists to determine where the rock was formed; the particles also record the periodic flip-flop of the Earth's poles, permitting aging of the rock. The results were astounding—the continents had indeed ambled and pirouetted all over the globe. Wegener had been right, and subsequent measurements around the world have given us a fairly good idea of how the continents were arranged in times past.[1]

One of the hardest things for a nongeologist to remember is that *here* wasn't always here. As I look over Linville Gorge—at the swallowtails drifting through the rangy crowns of the hemlocks and tiny spring azure butterflies that look like swatches of summer sky—everything looks so permanent. But these drifting islands of crust have wandered quite a distance. This little piece of the Appalachians, for instance, was once pivoted ninety degrees on its side and floating somewhere well below the equator. Same rock, different location. Combined with an enormity of time, the whole subject of geologic history becomes as slippery as an eel.

Nor have the continents always held their current shapes. As they drift, continental plates gather up the others' leavings, scraping up loose chunks and islands known as terranes in a process called accretion. Think of a hand sliding across a snowy windshield, piling up slush along its leading edge, and you get the general idea. Much of coastal New England and eastern Newfoundland, for instance, is apparently an accreted terrane, the remnants of a microcontinent which geologists named Avalonia.

The land never stays still for long; the continents dance across the mantle, bumping and grinding gracelessly, forming partnerships and dissolving them on a whim. The long, complex history of the Appalachians is rooted in this vast waltz.

In the briefest of nutshells, the Appalachians as we know them today were formed over a period of about 500 million years, in three separate mountain-building episodes, or orogenies. The first, the Taconic Orogeny, peaked about 440 million years ago and pushed up a chain of mountains that subsequently eroded to nothing, providing the material for the next wave of peaks. Then came the Acadian Orogeny, about 375 million years ago during the Devonian Period, the Age of Fish when the first amphibians were just deciding they liked dry land. These first two orogenies were largely responsible for what is now the northern Appalachians, while the third and most recent, the Alleghenian Orogeny, gave shape to the central and southern Appalachians about 290 million years ago.

If you take the history of the Earth as a calendar, with January 1 marking its formation, then the beginning of this sequence—the start of the Cambrian Period—doesn't come until the middle of November. The Taconic and Acadian Orogenies occur around the beginning of December, and the Alleghenian Orogeny happens on about December 10. Those who construct these geologic calendars always like to point out, with insufferable smugness, that we human party-crashers didn't arrive until a few moments before midnight on New Year's Eve.

The Taconic Orogeny came at a time when the continents looked much different than they do today. Africa, South America, Antarctica, Australia and parts of Asia were welded together in a supercontinent known as Gondwana. North America, Greenland, Ireland and Scotland comprised a land mass named Laurentia, which was tipped on its side from the modern position and separated from Gondwana by the Iapetus Ocean. (The names may seem confusing, but there is a method to the madness. For example, in Greek mythology Iapetus was the father of Atlas, just as the Iapetus Ocean was the forerunner of the Atlantic.) Finally, there was northern Europe, an island continent dubbed Baltica, which sat to the east of Laurentia.

Starting around 500 million years ago, Iapetus began to close, drawing Baltica toward Laurentia. The Baltica Plate slid beneath the Laurentian Plate, generating tremendous heat and sending lava plumes up through the coastal rock, causing the kind of volcanic mountains seen today along the Pacific Rim.

There is no mountain so tall that erosion cannot knock it down. By the end of the Silurian Period, about 400 million years ago, the mountains of the Taconic Orogeny were worn to gravel, which was deposited by rivers in a huge delta to the west that stretched from West Virginia to New York. These sediments eventually became the hard, erosion-resistant quartz sandstone and conglomerates that now cap many Appalachian ridges, and are exposed as the ghostly gray boulder fields of Pennsylvania's Kittatinny Ridge or the vertical knife-edge of Seneca Rocks in West Virginia.

The continents continued their slow-motion collisions. Laurentia and Baltica crunched together to form a new land mass known to geologists as the Old Red Sandstone Continent, in honor of a distinctively rusty rock strata. Where the two parts of this supercontinent met, a new range of mountains called the Caledonides rose, in an episode called the Arcadian Orogeny.

The Caledonide Mountains included what are now the northern Appalachians, but the range continued on the "European" side of the suture, forming mountains that marched across what would become Britain and Scandinavia as well as eastern Greenland. The Caledonides in what is now New Hampshire or Maine did not look like the modern-day White Mountains or Traveler Range. They were young and snaggletoothed then, like today's Alps. What we now see are the eroded stubs of once great peaks.

Later, when the Atlantic began to split open, the plates drifted apart, carrying the sibling hills thousands of miles from each other, but the haunted glens of Scotland's Highlands are close kin to the northern Appalachians. Little wonder that the Scots and Scots-Irish, forced to the New World by the brutal clearances of the eighteenth and early nineteenth centuries, found something familiar in Nova Scotia and North Carolina.

In the interlude between the Acadian Orogeny and the final mountain-building period a hundred million years later, tremendous changes took place. The land had become cloaked in forests—not of oak and hemlock, but of primitive seed-ferns, giant horsetails and tree clubmosses. Insects and other arthropods were abundant, and amphibians had aggressively colonized the land. With the Old Red Sandstone landmass sitting near the equator and shallow seas lapping both rims of the Acadian mountains, conditions were perfect for the growth of lush swamps.

The eventual result, of course, was coal—the product of millions of years' worth of fallen vegetation turned to peat in the oxygen-poor water, buried beneath later sediments, then compressed and heated into the varying states of coal. These ranged from crumbly lignite, or brown coal, through bituminous (soft coal), to the much altered, cleaner-burning anthracite, or hard coal. It's not surprising that this period of geologic history is known as the Carboniferous.

In the Appalachians, bituminous deposits are common in a wide swath from Alabama to western Pennsylvania and parts of the Atlantic Provinces in Canada. Anthracite, on the other hand, is restricted to a few narrow pockets in the Pennsylvania and Virginia ridges. I grew up in Pennsylvania's anthracite belt, a region scarred by the deep mines, with their outwash of stream-killing acid water, and huge strippings that gape empty-mouthed beside spoil banks.

The anthracite industry boomed in the nineteenth century but went into a precipitous crash in the early twentieth as other, more convenient fuels like oil dominated the market. More and more of the great breakers, which sorted coal from worthless shale and graded the chunks by size, squatted outside the towns, abandoned, the huge banks of windows pocked with smashed panes, the spindly conveyor belts that spread out like the legs of insects rusted in neglect.

The shale banks were unstable, dangerous places for kids, but I spent more hours than I could count combing their slopes for fossils. The coal swamps left their mark in the soot-gray sediments that the mining companies discarded, and few fossils from any time in history can match the elegance of those from the Carboniferous Period. I still know of no feeling quite so charged as splitting a slab of shale open along its weakest bedding plane and exposing the secrets inside. The seed-ferns in particular are exquisite, the imprints of their fronds smooth and glossy against the flat, fine-grained shale; the ancient leaves are sometimes still touched with a film of white or orange.

Many formerly enormous plants of the coal age have survived by becoming smaller. I remember the shock of recognition one day when I found a wonderfully preserved horsetail stem at the base of a shale bank, just feet from a feathery stand of living horsetails known as scouring rush. The modern species are scarcely two feet tall, but the Carboniferous versions were mammoth, reaching sixty feet or more into the air.

For more than 200 million years, the continents had been drawing closer and closer together, fusing in larger and larger clots. Now, as the Carboniferous Period came toward its end, only two huge landmasses remained, the Old Red Sandstone Continent poised in the north and Gondwana in the south—and they, too, were approaching each other, squeezing the Iapetus Ocean out of existence between them.

Their collision caused the last of the three Appalachian orogenies, named the Alleghenian by geologists. Western Africa smacked into the southern rim of the Old Red Sandstone Continent, forcing up the southern Appalachians. The impact, which stretched over a 50-million-year period, was epic in its force, reflected still in the tortured strata of the central and southern Appalachians and the plateaus of their western slopes. Across Pennsylvania the resulting ridges make a serpentine curve—the "fingerprint," in a sense, of Africa's bulge.[2]

Rock layers that had been deposited horizontally were folded, cracked and twisted, and much of the eastern rim of the continent was shoved as far as 160 miles to the west, over younger formations. Another 200 million years of erosion has, in some places, cut through weak points in the older rock, forming "windows" to the younger rock entombed beneath. Like Linville Falls, the famous valley of Cades Cove in Great Smoky Mountains National Park is a window.

Interstate highway road cuts are good places to see the results of the Alleghenian Orogeny's prodigious energy. In Pennsylvania, the builders of I-81 removed a piece of Spring Mountain, exposing the swirled layers of Carboniferous sandstone inside. Down the middle of each wall, however, is a diagonal slice–the Spring Mountain Thrust Fault, along which the rocks had been pushed some fifty feet from their original position. The Spring Mountain fault is a blip, however, compared with the Pine Mountain Overthrust on the Cumberland Plateau of Tennessee and Kentucky, a feature so big it can be seen from space. Here, as the giant piece of continental rim slid west, its southern flank hung up on a pocket of more stubborn rock; the force sheared off along the Jacksboro Fault, forming two distinct right-angle bends in the mountain. Satellite pictures show it clearly, but you can see its shadow on highway maps. With your finger, follow I-75 northwest from Knoxville. At Lake City, the overriding rock hit its snag at Walden Ridge, and the highway from here to Buckeye traces the fault line. Then the mountain, and the highway that follows it, make a precise ninety-degree turn to the right, back along the edge of the old crustal plate.

With the collision of Old Red Sandstone and Gondwana, the world's continents were finally united into a single entity: *Pangaea*, from the Greek word for "all land," which was surrounded by a single ocean, *Panthalassa*, or "all sea." Pangaea did not last, however; as it drifted north, it began to split down its middle, along the Iapetus rift that had only recently closed. The Atlantic Ocean was born, and over the next 200 million years Pangaea would tear itself apart as crustal plates went wandering in their own directions, a process that continues—albeit slowly—today. Europe is already fractionally farther from the Appalachians than it was when you started reading this chapter.

Rise and fall, uplift and erosion, compression and separation—squeezed into a few paragraphs, the cycle is neat, if dizzying. Then I look out the window at the Kittatinny Ridge and shiver at how long it takes just to wear a mountain down, to say nothing of doing it three times. Each new phase fed off the previous ones, recycling the sediments and, more often than not, leaving the foundations of the great mountains behind as reminders.

That's what today's Appalachians are, really—leftovers. Erosion's hand is everywhere, sculpting the violence of continental collisions into a landscape of subtle beauty. And because all rock is not created equal—some layers are more resistant to weathering than others—the work still goes on. The central Appalachian ridges, for instance, exist because the valley shales are softer than the mountain sandstones; this was a flat plain that became hilly not because the ridges rose, but because the valleys dropped.

Along the Virginia–West Virginia border, through the Maryland panhandle, and into south-central Pennsylvania lie thick, scattered deposits of Devonian shale. Where the hillsides face south, the shale beds are dry and blisteringly hot at ground level, up to 140 degrees on a sunny summer noon, and the loose, shifting rock makes for an unstable growing medium. These patches are, not unreasonably, known as shale barrens.

Forest plants cannot get a toehold, but shale barrens endemics, plants found nowhere else in the world, seem to thrive under these conditions. They include mountain parsley; a unique species of evening primrose; and a nonclimbing clematis, as well as prickly pear, the only cactus native to the

Appalachians. Cat's-paw ragwort, a delicate, yellow-flowered member of the daisy family, exhibits typical adaptations to the hot, dry environment—the whitish upper surfaces of its leaves reflect sunlight, while the densely furred undersides reduce water loss and insulate the plant from reflected heat.

The most famous shale barrens inhabitant is Kate's Mountain clover, a small species with white flowers tinted yellow or pink. Like many plants of the shale barrens, the clover has a long taproot that extends through the dry, loose rock to the damper, cooler soil more than a foot below. Named for Kate's Mountain in West Virginia, where it was first found, it is most closely related to buffalo clover, a prairie plant. This is typical of several barrens plants, which represent far-flung relatives of more common species; the shale barrens evening primrose's closest relative occurs in the Southwest.

As if the harsh climate weren't enough to worry about, shale barrens plants must also contend with human actions. Because they are found in only in a few widely scattered locations, they are at risk from overcollecting, road construction, even grazing. A number of sites, including the shale barrens at Kate's Mountain, have been ruined.

Running the length of the Appalachians is an even stranger habitat, which traces its ancestry back to ancient geology and which, like the shale

The White Mountains of New Hampshire are part of the northern Appalachians, the largest and least geographically cohesive of the chain's five regions.

beds, goes by the name *barrens.* The serpentine barrens are so called not because of their shape (although some do twist along hills like snakes), but because they are underlain with serpentine, a very unusual rock. Originally part of the floor of the ancient Iapetus Ocean, serpentine rocks like peridotite and gabbro were pushed high and dry by the later continental collisions, and now occupy a narrow belt from Georgia to Newfoundland. When first exposed to air, the rock has a smooth, greenish luster checkered with light netting that resembles the skin of a snake, but it weathers quickly into orangish soil.

Serpentine rocks and the soils formed by them are rich in all the things a plant doesn't need—iron, nickel, magnesium and chromium, which can be deadly in high concentrations—and low in most crucial nutrients like nitrogen and calcium. Consequently, serpentine barrens tend to support plant communities more typical of arid grasslands than the moist Appalachian forests; in fact, some barrens plants have disjunct ranges, such as prairie dropseed, a grass otherwise found far to the west of the Appalachians.

Except for geologists and farmers, most people don't make the connection between what's under the ground and what grows on top of it. In a serpentine barren, however, the link is forced on you. This is especially true in the largest and most dramatic of all Appalachian barrens, the desolate Tablelands in western Newfoundland's Gros Morne National Park.

I cannot imagine a more dramatic contrast than the scene that confronts anyone driving along Highway 431 from South Arm, one of the deep fjords that make Gros Morne so spectacularly beautiful. To the right are the Lookout Hills, green with spruce and a blanket of shrubs, with a few grayish cliffs poking through. To the left, across Wallace Brook and not a half mile from this thriving forest, is nothing but barren rock—a seeming desert of orange that covers the enormous hulk of Table Mountain, an area of nearly fifty square miles. Except for a few threads of green that trace the outlines of small creeks, there is scarcely any sign of a living thing.

The Tablelands is one of five large serpentine outcroppings in Newfoundland's Long Range. Most are in the section of coast from the Lewis Hills to Gros Morne, with an isolated outcropping in the White Hills at the island's far northern tip. The barrens are an extraordinarily difficult place for a plant to survive, not only because the soil is toxic and nutrient-poor, but because it is subject to severe churning and heaving when it freezes, an annual grinding that can chew a plant's root system to shreds; not surprisingly, the mountain slopes are unstable, with slides and soil creep common. These are hardly garden spots, and virtually no animals live here except for birds and the odd caribou passing through.

Yet some plants manage to eke out a living. Balsam ragwort, a relative of the cat's-paw ragwort of the shale barrens, has managed to cope with the toxins and can tolerate up to 12 percent of its dry weight in magnesium—a concentration that would level most flowers. Even the common pitcher-plant, a species normally associated with bogs, has a niche in this near-desert, growing along the edges of spring seeps where subsurface water brings up a little calcium. By supplementing soil nourishment with a diet of insects trapped in its upright tubes, the pitcher-plant is able to augment the Tablelands' miserly offerings. Several other carnivorous plants, including sundews and butterwort, work the same trick on their environment.

Moving carefully, painstakingly, from bare rock to bare rock, I worked my way up one of the small creek beds that score the slopes of Table Mountain. From a distance I must have looked ludicrous—a grown man moving with elaborate caution in a field of stone, alternating between mincing steps and long jumps—but my reasons were sound. The plants of the serpentine barrens are exceedingly fragile; for all their ability to tolerate harsh weather and inclement soil chemistry, they can be killed with a footstep.

Like the alpine plants of many Appalachian peaks, the barrens inhabitants are so dwarfed and nondescript that they are easily dismissed. Yet they can live to great age, growing with incredible slowness. Common juniper, which in southerly areas can reach a height of ten or fifteen feet, may grow no more than a few inches tall here, spreading out in a gnarled fan with a trunk no thicker than a man's thumb. Yet researchers, counting the thickly packed growth rings of these junipers, have found some more than three hundred years old.

The trickle of water that rose from the ground and flowed away in the tiny stream was bordered by a narrow margin of flowers—pitcher-plants with their purplish, nodding blooms, graceful blue harebells, the yellow flares of ragwort. Beyond the spring the slope steepened, leading to a wide, flat valley that reached nearly to the top of Table Mountain. Although the stream flows down the valley floor, it was not responsible for cutting this huge gash in the hill. That was the work of a glacier—the final chapter in the shaping of the Appalachians, and the period still writ most vividly across the land. Its signature can be found on valley floors in New England and Newfoundland, among the trees of a southern cove forest in North Carolina and on the highest mountaintops in between: the ice age.

It was midsummer in the White Mountains, and the traffic was racing up and down the Franconia Notch Parkway, one of the prettiest stretches of highway in New England. To either side the mountains rose high and abrupt, but the valley (known as a notch in this part of the world) was wide and flat-bottomed.

That valley profile—a broad U—is the telltale mark of a glacier, for running water cuts a different pattern, cleaving a V as it goes. About twenty thousand years ago, a thick and unimaginably heavy river of ice, carrying an abrasive slurry of boulders and smashed rock on its belly, ground its way down this valley. It carved the rock like a child gouging its fingers through clay, greatly enlarging an existing river valley into this long, cathedral space edged with cliffs and capped with mountains.

Glacial valleys score the Appalachians from the White and Green mountains north to Newfoundland. The landscape the glaciers created is among the most magnificent in the East, alarming in its sweep even if you do not try to fill it, in your mind's eye, with a mile-thick flow of ice. Glacial valleys like Franconia Notch or Crawford Notch are only the most dramatic examples of what the glaciers could do. Where the sheets overrode mountains, they sometimes cracked away the far sides, forming sheer cliffs; in other spots, like the high peaks of the Presidential Range, the alpine glaciers nibbled backward even as they flowed down, chewing bowl-shaped valleys known as *cirques*. Perhaps the most famous are the cirques of Mount Washington— among them, the Great Gulf with its fifteen-hundred-foot walls and Tuckerman's Ravine, where the snow lingers almost all year (just a little longer and the glacier would be reborn).

Not every New England notch is the direct result of glacial carving. In Vermont, about fifty miles northwest of Franconia Notch, is the smaller, less-frequently visited Smuggler's Notch, which has been a hiding place and clandestine travel route for contraband since at least the late eighteenth century; it even served runaway slaves heading for Canada during the Underground Railroad years.

Smuggler's Notch isn't a wide, U-shaped valley. Instead the walls cramp inward, narrowing to a tight V, and the valley floor has more twists than a blacksnake, with huge boulders blocking the way at every turn. A road goes through the notch—nominally two lanes, but it's polite to wait for the oncoming traffic to pass before you attempt some of the sharper hairpins.

Smuggler's Notch is thought to have been eroded by water—the great rush of melting glaciers, which accumulated behind a high saddle that sat between Mount Mansfield to the west and Spruce Peak to the east. Breaching this barrier, the water chewed down through the saddle, forming the torturous notch—and the perfect hideaway for anyone with illicit cargo.

The namesake smugglers are gone now, but the notch has reclaimed some of its former glory. The soaring cliffs are again home to peregrine falcons, which nested here until the 1960s, when they—like peregrines across the eastern United States—succumbed to the widespread effects of pesticide poisoning. In the 1970s, the privately supported Peregrine Fund began releasing captive-bred young falcons in the East, a program that came to the Green Mountains in 1977. Peregrines usually nest on rock ledges at dizzying heights; in Smuggler's Notch, they've picked a spot near the precipice known as Elephant's Head. A pair set up territory here in 1987 and raised chicks for the first time two years later.

At the south entrance to the notch, I stood along the road and looked up at the cliffs. Somewhere in the birches and maples, across the road and up the slope, a mourning warbler was singing, but much as I would have liked to look for it, I couldn't. The whole area was posted with signs against entering: "Restricted Area. Falcon Nesting Area. $1000 Fine for Disturbing." The warbler would have to wait.

A bird soared into the open air above the gorge, wheeling, but it was only a raven; these huge corvids nested in Smuggler's Notch during the peregrine-less decades, and I found myself wondering how the return of such an implacable aerial hunter affected them. Once, at a falcon nest in Maine, I saw a raven flapping lazily along the spine of a mountain about a quarter mile from the eyrie. With no warning—for me or the raven—a peregrine sizzled down from the sky in the classic, hurtling dive known as a stoop, reputed to reach two hundred miles per hour. A stooping peregrine can kill birds much larger than itself with the sheer force of the blow, but this raven tumbled frantically aside at the last possible second, narrowly avoiding such a fate. Thin and distant I heard the enraged *kek-kek-kek-kek* of the falcon, as the raven beat a furious retreat, croaking its displeasure.

The raven was gone, and the skies above Smuggler's Notch were empty for a heartbeat. Then I saw two silhouettes coming from the east, riding into the headwind with the characteristic crossbow shape of a peregrine—wings curving back gracefully from the heavy, muscular chest and the blunt head. One caught a thermal and soared, making three full circles, while the other folded and dropped, losing itself against the cluttered background of the cliff. When I looked again, the first falcon had nearly doubled its height, lofting on the column of rising air; then it set its wings and continued west, no doubt intending to hunt. Somewhere up on

Elephant's Head was a ledge with two ravenous chicks, begging for food where the waters of a glacial torrent once ran.

So far as we can tell, the Earth has experienced five major glacial ages within its 4.6-billion-year history, three within the past billion years. As with so much of the planet's history, the underlying causes, while complex, are at least partially tectonic. When the continents drift over the Earth's surface, they sometimes arrange themselves in patterns that inhibit the movement of warm, tropical seawater to higher, colder latitudes. This happened most recently about 20 million years ago, during the Miocene Epoch. Australia spun away from Antarctica, deflecting warm ocean currents toward the South Pole, where Antarctica itself fetched up. In the north, Europe, Africa and Asia coalesced into a huge continental complex, pinching off the Arctic Ocean from most sources of warmer, mid-latitude water, while the Isthmus of Panama rose and blocked currents trading between the Atlantic and Pacific.

The effects of these continental movements were profound. With a continent over one pole and a shallow, largely landlocked ocean over the other, there was no easy way for warm, equatorial water to transfer its energy to the cold, high latitudes. On top of that, most of the world's land masses were congregated high in the northern hemisphere, far from the equator; such a placement invites an ice age.

Still, the actual development of huge ice sheets requires a trigger. As early as the 1860s, a few scientists speculated that there were very long-term cycles in solar radiation that could bring on ice ages, a theory polished by Yugoslav mathematician Milutin Milankovitch in the 1940s; these cycles bear his name today. At their most fundamental, there are three minor cycles (in the tilt of the Earth's axis, the shape of its orbit and the time of year when it is closest to the sun) that produce an overall, one-hundred-thousand-year Milankovitch cycle of glaciation.

The story of the ice ages is the story of cycles within cycles. During that one-hundred-thousand-year rotation, the glaciers grow for about 60 to 90 percent of the time, and shrink for about 40 to 10 percent; the decline is generally more sudden and precipitous than the increase. The periods of growth are known as glaciations; the warmer interludes are called interglacials. Within the current glacial age, which began about 3.5 million years ago, there have been nineteen or twenty expansions and contractions of the ice sheets. Each has been named by geologists, and the most recent is known as the Wisconsin glaciation in eastern and central North America. It started some seventy-five thousand years ago and reached a peak roughly twenty thousand years ago. The interglacial that followed it—the one we are currently enjoying—comprises the Holocene Epoch.

At their height, the Wisconsin glaciers covered the northern half to two-thirds of North America, from the High Arctic to Newfoundland and south to Pennsylvania, the Great Lakes (themselves gouged by ice) and across the northern edges of the United States. There were two main ice sheets—the Laurentide, centered on what is now Hudson Bay, and the much smaller Cordilleran sheet growing from the mountains of western Canada and the Alaskan panhandle. The two sheets may or may not have met—a question of great importance to the timing of human arrival in North America, as we'll see later.

The sheets were more than three miles thick in places, and they locked up much of the world's water within them. Sea levels, consequently, were as much as three hundred to five hundred feet lower than they are today, and the continental shelf now submerged off the Atlantic Coast was high and dry. The Grand Banks, the Georges Banks—many of the fertile fishing grounds of our age—were home to herds of grazing animals such as mammoths. The proof comes from fishing trawlers, which have dredged up enormous mammoth teeth with their catch.

In order for a glacier to grow, all that is needed is for more snow to fall in winter than can melt in summer. As the snowpack accumulates it solidifies into ice, expanding outward from its middle as more and more snow builds up in the central dome; the process is a little like pouring thick pancake batter onto a griddle. But what a griddle! At its peak, the Wisconsin glaciation had some 15 million square miles under its thumb.

Looking at a map of North America during the glacial maximum, with ice blotting out nearly two-thirds of the continent, one finds it hard to believe that life could have survived. But it did, rather handily. True, plants and animals had to retreat south in advance of the glaciers, and undoubtedly some species (plants in particular) were driven to extinction. But many seem to have bided their time in southern or coastal refuges, away from the ice, ready to return with the thaw.

One of the great ironies of the past hundred thousand years, in fact, is that life appears to have had an easier time dealing with the glacial maximum than with the balmy interglacial that followed. When the glaciers were at their peak, North America boasted an unprecedented array of large mammals—the ice age megafauna, as paleontologists have dubbed them. "Mega" is right. They included two elephants, Jefferson's mammoth and the American mastodon; giant ground sloths that stood twenty feet tall; and short-faced bears several times the size of modern grizzlies. There were dozens more—long-horned bison, stag-moose, shrub-oxen that looked like a cross between yaks and bighorn sheep, huge dire wolves, American lions twice as big as African lions, saber-toothed cats, woodland musk ox, several species of pronghorn, American cheetahs, camels and horses. Although there is uncertainty about their precise ranges, many of these giants were Appalachian.

The perplexing twist on this story is not that all these strange and wonderful species became extinct—but that they became extinct well after the glaciation ended and the interglacial hit its stride, about twelve thousand years ago. The mass dying claimed nearly fifty kinds of mammals, almost all of them megafauna. Giant wolves vanished; smaller canids such as gray and red wolves and coyotes survived. American lions and sabertooths winked

A trout lily, its yellow petals arching back to form an airy globe, blooms in the warm sun of early spring.

out; smaller mountain lions and lynxes did not. Cheetahs and all but one species of pronghorn became extinct; it is because they once had to flee fifty-mile-per-hour cats that today's pronghorns can still run so fast.

Because the disappearance of the ice age megafauna coincides rather neatly with the accepted arrival date for humans in North America, a persuasive school of thought—the Pleistocene overkill theory—blames the disappearances on excessive hunting. Proponents point out that many of the survivors, such as moose, bison and caribou, were themselves immigrants from Asia across the Beringia land bridge; these animals, so the argument goes, were already adept at avoiding human hunters. North American natives like mastodons were not, and were exterminated for their ignorance, taking their predators along with them.

Others claim that the extinctions were triggered by the drastic changes in climate and habitat as the ice sheet shrank. The ten thousand years during which the glaciers were shrinking saw a northward expansion of plant and animal communities from their ice-free refuges in the south and along the coasts, and a sometimes rapid replacement of one habitat type with another. This may explain the extinction of American mastodons, which thrived in the vast forests of white spruce growing in the immediate wake of the glaciers. According to fossil pollen records bored out of lake bottoms, the spruces experienced a massive, sudden collapse across almost all of their range about eight thousand years ago. The spruces eventually recovered somewhat, but not so the mastodons.

What would the Appalachians have looked like at the very end of the Wisconsin glaciation? Nearly half the range, from Newfoundland south to northern Pennsylvania, would have been engulfed in ice, with only a few ice-free refuges where the highest mountains protruded, or where the lobes of the ice sheet skipped small sections of coastline. These alpine *nunataks*, as they are known, are thought to have occurred in the Long Range Mountains of Newfoundland, on Cape Breton Island and in the Shickshocks of Quebec, as well as on such New England mountains as Katahdin and Mount Washington. Small coastal refuges are theorized for the west coast of Newfoundland and the tip of the Gaspé Peninsula.

Along a narrow margin just south of the Laurentide ice sheet (and to the east, along the exposed continental shelf) grew tundra, the dwarfed community of woody and herbaceous plants that can scratch out a living in conditions of cold, wind and almost perpetually frozen soil. This tundra zone, which from pollen samples appears to have been unusually grassy by modern standards, also extended south in a tongue along the Appalachians through West Virginia, Virginia and the Smokies.

Next came the widest vegetative zone of all, the boreal forest—the kind of conifer-dominated ecosystem now found across much of Canada. Fairly open, park-like boreal forest of jack pine, spruce and fir would have covered most of the midsection of the continent, from the margins of the tundra zone in Pennsylvania well down to about the middle of Georgia. Hardwood forests of oak and hickory (mixed with pines), which are much more susceptible to cold and drying than boreal conifers, were restricted to what we think of today as the Deep South; until the climate began to warm again, they were banished from the Appalachians altogether. And the species-rich mixed mesophytic forest, which reaches its greatest expression as the modern cove forests, was confined to moist river valleys within the hardwood zone.

After the great freeze came the great thaw. Continental glaciers die much more rapidly than they grow, and the demise of the Wisconsin ice sheets is thought to have occurred in just ten thousand years, as the Milankovitch cycle moved the Earth into a slightly more temperate attitude. Glaciers die not only from warmth but also from starvation, and this one-two punch is what is thought to have doomed the most recent incarnation of ice. First, warmth nibbles away at the glacier, freeing fresh water that begins to raise ocean levels and float the margins of the sheet. The runoff forms a lens of cold freshwater on the denser ocean water, reducing evaporation and, with it, precipitation. When snowfall drops below the critical level needed to maintain the glacier's growth, the retreat is sudden.

By about 14,000 years B.P. (before present) the ice was melting much faster than it could be replenished, and the margins of the glaciers were being eroded north at an accelerating pace. At first this merely widened the tundra zone, for few trees can survive the frost-churned, saturated soil that a decaying glacier leaves in its wake. But over the next four thousand years, up to about 10,000 B.P., the ice sheet retreated to the Canadian border, drawing the plant communities behind it like a rug. Tundra was restricted to the Maritimes and Newfoundland, with boreal forest no farther south than the Mason-Dixon line. Hardwoods had reclaimed the southern Appalachians.

But the warmth didn't stop there. The apogee of the Milankovitch cycle, known as the *hypsithermal* (from the Greek "highest heat"), came around 7,000 B.P. in the Appalachians, when conditions were actually warmer and drier than they are now. In places, the Arctic tree line was nearly 175 miles farther north than it is today, and many species of plants (and presumably animals) were also found north of their modern ranges. A few still remain, like refugees, such as the scattered rhododendron stands in northern New

England or the black gum swamps of southern Vermont. In western Pennsylvania, on the rolling Appalachian Plateau, lies a small state park known as the Jennings Environmental Education Center, which preserves a tiny window on the hypsithermal. It is a relict prairie, a last surviving fragment of a grassland tongue that during the prolonged warmth reached east from the Plains into the hardwood forests of the Appalachians' western slope.

Each summer, this miniature prairie explodes with color—purple columns of blazing star flowers, yellow masses of goldenrod, pale pink from wild spiraea. A layer of clay and fine glacial silt acts like a pool liner to keep the ground wet and the forest at bay, a reminder of a warmer, grassier past.

The hypsithermal is history—or ought to be. The climate should be on a downward track again, rolling toward the next glaciation, and for a while it was. There were blips, of course—the upward blips like the moderation known as the Little Climatic Optimum, which brought the Vikings to Greenland and Newfoundland a thousand years ago, and downward blips like the Little Ice Age, which chased the Norsemen back out again a few centuries later.

All things being equal, though, the long, long-range forecast should be for ice. That's how the glacial cycle has worked for millions of years. But things aren't equal. A little carbon dioxide in the atmosphere goes a long way, as scientists have known since the nineteenth century, and since the Industrial Revolution we've spewed far more than a little of it into the air, along with a lot of other greenhouse gases.

In ways that are becoming more evident each year, we have altered the very rhythms by which the planet has existed for the past several million years, short-circuiting the cycle of freeze and thaw that has dominated life for eons. The Pleistocene is gone, and with it the ice and the great mammals. Gone now, too, is the Holocene, the natural breathing space of warmth between glacial advances.

We are, a growing number of scientists agree, in a wholly new and unprecedented epoch in Earth's history: the Anthropocene, a time when humans have perturbed virtually every aspect of the planet's natural systems, from climate to ocean chemistry to terrestrial ecosystems. How long this may last—how long we will last—is unclear, but with every fresh ton of carbon we release into the air, the future becomes inexorably warmer. Until the Anthropocene and its makers pass, the days of the glaciers seem to be gone for good.

Sharp-shinned Hawk

Chapter 2

ISLANDS IN THE SKY, TRAVELERS ON THE WIND

Sharp and cold as broken ice, a November north wind was scouring the boulder field, setting into nervous motion the bare twigs of stunted chestnut oaks and red maples that grew at the forest's edge. At my back, a mountain ash shivered with me in the chill, while its clusters of orange berries hissed against each other by my ear.

The air was brittle and exceedingly clear; from my perch atop the Kittatinny Ridge, I could look northeast to the sudden rise of the Pocono Plateau or south, across the Pennsylvania Dutch farms and beyond the jutting edge of the Pinnacle, where the Kittatinny makes the second of two confused, hairpin turns before resuming its sinuous course to the Maryland border, 120 miles to the southwest.

I sat as low as I could, hands jammed in my pockets and my ears smarting in the cold, trying to shrink behind the shelter of an upturned slab of rock. It was no use; the wind was everywhere. I stood, painfully, to go.

As I did, a flicker of movement caught my eye. A few yards back, a small bird had landed on an outcropping of gray sandstone, a creature as pale as the rock itself. The size of a sparrow, it was white with a crown of faded buff and a mix of white, black and brown on its wings. The bill was short and conical, the color of dead grass; the eye was perfectly round and black, with an active gleam.

It was a snow bunting, chased south by the onset of winter from the High Arctic, where it had nested. A minute later, several more fluttered down from the sky on white-marked wings and joined the first, hopping among the jumbled boulders and looking for the last of the tree crickets that had lived all summer on the rock field. A bunting reached into a crevice and extracted one of the soft, pale green insects, so numbed with cold that it barely moved as it was eaten.

Buntings are open-country birds, residents by summer of the tundra but which move in winter to farm fields and sea dunes across southern Canada and the northern United States. They are not mountain animals—except for a few weeks in the fall, when the Appalachians provide many of them with a convenient pathway to the south, where a handful winter on the highest grass balds. The late Maurice Brooks, dean of Appalachian naturalists, looked on these small, white birds as potent symbols of the biological cohesiveness of the mountain chain.

The buntings are not alone in this hegira along the twisting ridges of the Appalachians. An astonishing variety of species—most of them birds, but including even insects—use this remarkable skyway as a funnel from north to south and back again. It is a community of travelers largely unknown just a thousand feet lower, drawn from the farthest reaches of the continent's north: monarch butterflies from milkweed fields in New York; dragonflies from the glacial bogs of the Poconos and New England; loons

Necks craned and binoculars ready, hawkwatchers scan the sky for migrants at Hawk Mountain Sanctuary, along Pennsylvania's Kittatinny Ridge.

from Canadian lakes; saw-whet owls from New Hampshire; tundra swans from the Northwest Territories; and golden eagles hatched in the remote Ungava on cliffs that overlook Inuit villages. Also among its members are pine grosbeaks and red-breasted nuthatches from the boreal forests—and the sharp-shinned hawks and merlins that eat them. Like a river they flow over and through, heading for heaven only knows where.

That they follow the Appalachians should be no surprise, for the mountain chain is nearly perfect as a migratory route. Most importantly, it is long and oriented north to south, bridging more than two thousand miles from Quebec and Newfoundland to Alabama. Just as critically, the mountains are continuous, so migrants do not have to contend with wide areas of open valley. A lack of such continuity makes the Rockies much less important as a migratory pathway, for to use them birds must leapfrog from one disjunct mountain system to another.

As if that weren't enough, the Appalachians' central ridge and valley province, with its evenly spaced, parallel hills, is just right for migrants, like lanes on a superhighway. As one drives north across the Great Valley of southeastern Pennsylvania, with South Mountain (the most northerly prong of the Blue Ridge) fading in the distance behind, the first of the ridges rises above the trees to become a long, horizon-hugging hill. Along most of its length it is known as Blue or First or Broad Mountain, but students of natural history prefer its Lenape Indian name: Kittatinny, or "Endless Mountain."

The Kittatinny Ridge's prominence dates to the 1930s, when word of great hawk shoots along its summits leaked out. Photographs of piles of dead hawks spurred a small group of conservationists to buy the mountaintop where the worst of the slaughter was occurring. They declared the parcel Hawk Mountain Sanctuary, the world's first refuge for birds of prey. That was in 1934, when few people realized that the Kittatinny is one of the world's greatest migratory pathways; today, the parade of hawks, eagles and falcons drifting down the ridge draws hundreds of thousands of people to Hawk Mountain's North Lookout and dozens of other hawk-watches in the Appalachians.

When I was twelve, on a blustery October day in which the sky was tattered with clouds, a sharp-shinned hawk folded its wings into a teardrop and dove from the air straight toward where I sat on the boulders of North Lookout. In my cheap binoculars, the bird swelled and grew larger, until it filled the glasses and I dropped them in shock. The small hawk flashed a few feet over my head and did a hard right bank behind me, whacking its talons on a plastic owl decoy that sat on a pole jammed in the rocks.

For days thereafter, whenever I closed my eyes, I could see the slim bird expanding in my view, its cranberry-red eyes, the slate-blue back, the rusty bars across the breast—the mastery of wind and gravity and cold animosity that brought it like a lance from the air to the owl, and almost within my breathless reach.

I had seen hawks before, but never with such clarity or with such terrible intimacy. I was hooked.

I grew up in the shadow of the Kittatinny, and since that day its bouldered spine has been the focus of my world; my roots are not home or family, but the Tuscarora sandstone of the mountain, and my season is the autumn. I am not alone in this, for there are many others besotted with hawks. When most people are luxuriating poolside in the sticky oven of a Labor Day heat wave, we watch the weather maps for a bulging line of cold air, sagging down out of the Canadian prairies, pregnant with wind and hawks and the smell of fall.

The migration begins in mid-August, after a night of lightning and crashing wind. The cool aftermath of the frontal passage brings the vanguard of bald eagles and ospreys and kestrels, a trickle that becomes a flood in the weeks that follow. Broad-winged hawks by the tens of thousands fill the September skies; in October, sharp-shinned hawks and Cooper's hawks hedge-hop down the ridge, skimming the canopies of the bronze oaks. Each month has a different flavor—peregrine falcons in October's early weeks, red-tailed hawks on the bitter gales of November, the last golden eagles as the migration dies in December's snow.

One autumn, I set everything else aside save for the mountain and the hawks. Every rainless day for three months, I helped run a banding station a few miles upridge from Hawk Mountain, culling raptors from the sky with nets and deceit, then sending each on its way with a shiny metal ring joined loosely around its leg. From my windswept perch, I watched the cornfields far below change from green to ocher as August blended into September, saw the maples in the stream valleys flame to crimson in October and then watched the orange of the oaks creep down the flanks of the hills as the leaves turned. Some mornings the fog hung thick in the valleys, marooning us in the sunshine above. Over time (and perhaps because of those foggy dawns), my perspective about the ridge changed in ways that I had difficulty articulating, even to myself. I no longer saw it as a valley dweller sees a mountain, distant and dark, a blue hem upon the horizon. Instead, peering down from its top day after day, I started seeing it as an island of sorts, floating over fields and villages that now seemed remote and unreal. It may be that I had finally learned to see the valleys as a hawk sees them.

Drawn from nesting grounds in Greenland and the Canadian Arctic, peregrine falcons follow the Appalachians south each autumn, part of an immense, seasonal river of raptors.

October 8

Torrential rains all afternoon. In the meadow by my house, the small stream is over its banks, a red-brown smear of silt carried down from the neighbor's pasture. At sunset the clouds tear loose from the horizon in a ragged slit, and a horizontal shaft of orange light illumines the hills for a moment. To the east, a rainbow arches; then the sun disappears and it is gone.

An hour after dark, the rains cease completely and the wind begins to howl. This is an elemental wind, savage and cold, shaking the house with its hammer blows; the treetops surge like kelp confused by an angry surf. I stand outside under vivid stars, watching fragmented clouds race beneath them, leaning slightly into the blast as leaves are ripped from the trees overhead and cascade around me like froth in a cataract. Unconsciously, I stretch my arms overhead, reveling in the wind, feeling fey and pagan, wanting to dance with the leaves. There will be little sleep tonight, for my dreams will be haunted with the promise of wings.

The mysteries of the fall migration drew us, a small group of researchers and a few amateurs like myself, hoping to crack some of the riddles posed by the annual passage through the Appalachians.

The trouble with studying wild animals is that each member of a given species looks pretty much like all the others. For someone who simply enjoys watching wildlife, this is of no consequence, but for a biologist, it is maddening. The Appalachian hawk migration is an excellent example; for more than eighty years, ornithologists have taken systematic counts at hundreds of sites from North Carolina to New England, recording such minutiae as flight altitude, species composition, wind speed and direction, cloud cover—even whether the hawks had fed that day, based on the size of their crops.

One might expect that such long-term study would have answered the fundamental questions about hawk migration, but that is not the case, for mere observation cannot discriminate between individuals. There is no way of knowing if the broad-winged hawk that flies past Sunrise Mountain in New Jersey is the same one that passed Bald Mountain, Connecticut, the day before or will sail past Waggoner's Gap in Pennsylvania the next day. There is no way of knowing where they have come from or where they are going.

In order to trace a bird's particular wanderings, a biologist must be able to identify individuals. The easiest way is by using numbered bands, color tags or tiny transmitters—and such techniques obviously require a hawk in the hand. So each day we set our traps—mist nets as fine as hair and tough as rope, invisible against the edge of the trees; heavy bow nets powered by springs and concealed on the ground; wire cages containing pigeons and festooned with monofilament nooses to ensnare a hawk's toes. In a camouflaged blind, we sat in silence, watching the sky to the northeast for the specks that slowly resolved themselves into hawks.

Out front in the clearing, a pigeon sat in the middle of the bow net, wrapped in a protective jacket of heavy leather and attached to a cord that led up a tall pole and back to the blind; with a practiced pull, we could make the bird flap high into the air and land, once again, in the middle of the hidden net. We waited.

October 9

The wind keens all day, tearing the leaves from the red maples and flinging them into the mist nets, where they hang like drops of blood in a spider's web. The lure bird, a white pigeon named Snow, lofts easily into the wind when Robert pulls on the cord, and lands with its head up, nervous. Taking the hint, I look around, but there is nothing. Then, just that suddenly, there is something; the scudding, gray-bellied clouds have birthed a torpedo, low and near and closing fast. I have only a second or two to register the image—a Cooper's hawk as big as a crow; wings tucked, legs crooked, long, yellow talons drooping in misleading innocence. Head-on, the hawk lacks the grace one expects from a raptor, and in the instant before it hits, I see the bird as its prey must see it—naked in its threat.

The strike is a blur; two feet from the pigeon, the hawk throws open its wings and slams its fanned tail against the air, braking. The pigeon darts right, but the hawk pivots neatly above it even as it lashes down with its talons, pinning its prey to the ground.

Then follows a second's utter stillness, the hawk standing atop the pigeon with one wing thrown wide for balance. Out of the corner of my eye, I see Robert's arm convulse as he yanks the trigger handle, and the hidden bow net scythes through the air over the hawk.

Each of the hundreds of hawks we caught became, albeit unwillingly, an object of science. Each was carefully measured and weighed, its crop gently felt for recent meals, the bulge of fat beneath the wing and the thickness of the breast tissue checked as a general indication of health. The old, worn feathers of the wings and tail contrasted with their glossy replacements, and we noted the molt sequence carefully. Blood samples were taken for analysis; some birds were tagged with thin tail streamers, in the hope that observers downridge would record the date and time of their passage.

Most importantly, around each hawk's right leg was loosely crimped a metal, lock-on band with a sequential number and the words AVISE BIRD

BAND WASH DC. If the hawk were ever recaptured (or, more likely, found dead) and if the band number were reported to the U.S. Geological Survey's bird banding lab, one more tiny chink in our knowledge of bird life and travels would be plugged. Since fewer than 3 percent of the banded birds are recovered, banders must play the odds; most birds simply sink into anonymity the instant they leave our hands.

October 15

The sky blackens in the west and showers float over us, blurring the hills in gray. Through the drizzle comes a single, hard-edged shape, black against the rain, revealing flashes of white in the gloom. The eagle is pumping low along the north side of the ridge, appearing and disappearing in the wraiths of mist. From more than a mile away, it spots the pigeon flapping against the darkness of the woods and arrows toward us; I have never seen an eagle so committed to coming in. Like a redtail it drops down to the treetops, slipping among them for cover. With a shock I realize there are actually two adult bald eagles coming in, nose-to-tail like fighter jets. They cross the clearing just fifteen yards apart and drop almost to the ground, skimming in knee-high at maybe forty miles an hour. At such speed, at such close range, they look vast and frightening.

In these last critical moments, Bill keeps the pigeon in the middle of the bow net, gripping the lines tightly to prevent the eagle from snatching the lure out of the safe zone. The pigeon, seeing what it presumes to be its doom descending upon it, gives a frantic flutter. This image—eagles, pigeon, the soft rain slanting from the north—fixes itself in my mind.

The lead eagle pulls up within spitting distance, swirling the long grass in its slipstream. The other swerves right. Then both are gone, and no one speaks for a long moment.

In the decades since people began to consider raptors as something more than gunnery targets, we have learned much about the grand passage each autumn down the Kittatinny and the other ridges of the Appalachians. We know, for instance, that more than sixteen species of hawks, eagles, falcons and vultures use the mountain chain as a flyway. For some, the journey is short; a red-tailed hawk hatched in eastern New York may slide off the ridge in central Maryland, content to spend the winter in rolling horse pastures. Others of the same species, for reasons we do not understand, may push onward to the coastal plain of North Carolina or the flat prairies of south Florida. Others go much farther. A broad-winged hawk from the Maine woods may face a trip of more than four thousand miles, ending in the jungles of Peru. Peregrine falcons from Greenland and Labrador may use the ridges as a convenient route on their way to the Mexican coast and Central or South America.

We know that the northwest winds of autumn are pivotal to the migration, striking the Appalachians at a roughly ninety-degree angle and providing deflection currents that give the hawks lift; on days of little or no wind, the raptors may not fly at all, or they may strike out due south across the valleys, skipping from thermal column to thermal column like sailplane pilots.

But there is much we do not know. How does a young falcon, flying alone on its first trip south, know that the ribbon of hills beneath its pointed wings will lead it to a warm, safe place to pass the winter? The falcon isn't saying.

October 16

Lunchtime, and I'm holding the pigeon line with one hand and eating a sandwich with the other, only occasionally laying the food down to give the lure a twitch and scan the sky with binoculars. While making a perfunctory sweep, I see a high bird, hanging against the clouds—long wings gracefully tapered, a long tail, with a hint of power in its frame. I try to say the word "peregrine" around a mouthful of bologna and Swiss, but it comes out garbled. Still, everyone else in the blind recognizes my message, and all eyes look heavenward as I start popping the pigeon.

"There it is!" someone hisses, and I long to look up, but I concentrate on the pigeon, trying to control my nerves. Peregrines are not just birds; they are the fastest of living things, among the rarest of raptors. A peregrine falcon is power personified, avian evolution's finest effort, and here along the ridge, catching one is like attaining the Grail.

The falcon drops almost vertically on a pigeon in a searing stoop, then pulls off at the last possible moment and towers up again. This is typical; the peregrine is trying to flush its prey, so it can make the kill in midair. Like a pendulum the falcon drops again from the opposite direction, scant inches above the pigeon, and hits the middle of the mist net—a textbook example of how the trap is supposed to work.

It is an immature bird—a female, judging by its large size, an Arctic peregrine for sure, since it wears the crown of polished, golden feathers that is characteristic of the tundra race. The back is chocolate, with a delicate bloom that has a faint bluish iridescence; lighter spots on the undersides of the wings are shell-pink. The feathers—even the breast feathers—look as hard as scaled armor, a streamlined masterwork, and I find myself gently brushing them, almost surprised at their softness. The falcon glowers at me with umber eyes that are limitless and dark, speaking of treeless horizons and the flash of sun on the Arctic sea.

Peregrines are the icing on the cake, but the bread-and-butter birds of the autumn flight are sharp-shinned hawks, Cooper's hawks, broad-winged hawks and red-tailed hawks. The sharpies (as they are known) and Cooper's are accipiters, agile forest raptors that hunt small birds and mammals; they are masters of ambush and the sneak attack, with round wings and a long, rudderish tail for snap turns in the tight quarters of the forest. For them, migration through the open air must feel out of character, for during the rest of the year, they are never far from cover.

Redtails, on the other hand, revel in the limitless sky. They are buteos, members of the clan of soaring hawks (with broad-winged hawks and red-shouldered hawks), whose fan-shaped tails and wide wings can coax the most fleeting bit of lift from a thermal. These are the hawks that scribe lazy circles in the hazy summer air—mammal hunters, for the most part, watching the alfalfa fields for the skittering brown shape of a meadow vole foolish enough to cross a mown strip. They prey on animals with teeth, and the redtails we caught often carried old bite wounds, scabby and brown, on their toes, or dried blood and bits of fur from a squirrel or rabbit on their talons.

The tempo of the fall flight always surges and ebbs from day to day; a cold front brings northwest winds and a heavy passage of hawks, then the winds switch to the southwest for the next day or two as the high pressure cell drifts off the coast, and the air becomes still but for a fitful breeze. The temperature rises, the number of hawks declines and the humans who sit in anticipation of their arrival may doze on the rocks below, waiting for the next Arctic blast.

November 11

Rain in the morning, but we take the weatherman at his word that the front will pass soon, and hike up to the station at lunchtime. The clouds are still low but rainless, a wet ceiling just a few hundred feet above our heads, whistling along on strengthening north winds. It is hideously cold, clammy as we are with sweat and the dampness of the air. A few redtails pass with a bald eagle as the winds increase even more, slicing through the barren trees; gusts slam against the canvas side of the blind, snapping it with a report like a whip's. The clouds clear, and the flight begins in earnest.

The redtails that we catch come down in blistering dives, sometimes almost from the limit of vision, under perfect control in winds blowing up to forty-five miles per hour. They crab in, matching their forward flight to the perpendicular element of the gusts like pilots landing in a crosswind. As they approach they slowly tuck their wings, gaining velocity as the dive steepens. Roughly seventy-five yards out, they fold their wings almost completely and become missiles. At the last moment, each one makes a sliding turn downwind, then pirouettes into the headwind just above ground level, at the precise instant it opens its wings and tail for braking. One after another through the day, with perfect control, they tear the sky open and leave us numbed by the beauty of it.

In November, the burgeoning migration of red-tailed hawks is the tapestry through which a grace note of gold is woven. Most bird-watchers get excited when a bald eagle flies by the lookout, but something about a golden eagle has always made my pulse race. The lines are smoother, cleaner, more predatory. The proportions are balanced between the length of the wings, the breadth of the tail, and the heft of the body, where powerful muscles hide beneath mahogany feathers.

Golden eagles in the East have always been a mystery. When Maurice Broun, the first curator at Hawk Mountain and the pioneer of hawk-watching in the Appalachians, began recording the fall count in 1934, he was amazed to find this finest of North American raptors passing by his lookout—and not just one or two that may have been blown east by a storm, but fifty or sixty a year.

The ornithological community scoffed at first, but when the eagles kept appearing year after year, even the most doubtful had to concede that a relict population of golden eagles had survived somewhere in the East, and still kept time with the seasons by migrating south through the Appalachians.

In the eighty years since then, our understanding of eastern goldens has grown only marginally. Nests were found in the most remote corners of the

Adirondacks and Maine, but those who knew details remained understandably tight-lipped, lest egg-collectors, pirate falconers or overzealous photographers disturb the pairs. Most Appalachian goldens come from eastern Canada, hatched in widely scattered eyries on cliffs above the boreal forest and along the tundra escarpments north of the tree line, from the Gaspé up through the desolate Ungava Peninsula between Hudson Bay and the Labrador Sea. While western goldens hunt mostly mammals such as ground squirrels and jackrabbits, their eastern counterparts are members of what has been called the "bittern ecosystem," dependent upon bogs and marshes, and feeding on muskrats and wading birds including bitterns and herons.

In winter they can be found here and there but never in great numbers: a few in the farm valleys below the Kittatinny or in Virginia, one or two keeping company with bald eagles on the lower Susquehanna or Maryland's Eastern Shore, others in the southern Appalachians on the high, alpine "balds" above tree line or on the milder coastal plain between the mountains and the sea. Most, though, move like ghosts, appearing only briefly before human eyes, then once more passing beyond our ken.

All but extinct in the East as recently as the 1980s, bald eagles have experienced a dramatic recovery, and the sight of immature birds like this one is now common along the Appalachians each autumn.

November 13

I have never experienced an eagle day like this one—fifteen in all, ten goldens and five adult balds. At first we are too busy to notice, for in the course of the day we catch thirteen redtails and four goshawks and have our hands full with the tedious work of processing each one. But by afternoon, it is clear that something remarkable is afoot. The eagles are coming through in bunches, two or three at a time; a few respond to our lures, although their stoops are half-hearted, sliding over for a look but pulling off without committing themselves.

At 3 P.M. we spot two adult bald eagles going by at eye level to the north. Mark pops the pigeon, and one of the birds veers off its course and flaps our way, staying high; in most cases, a high approach signals a lack of interest. In the late sun, the white head and tail shimmer like mirrors, and the black undersides swallow the light. The eagle hangs overhead about two hundred feet up, then pulls in its wings and parachutes gently down, lowering legs that look as thick as a small child's, with brilliant yellow feet. Down and down it drifts, until it changes its mind and flaps for height, sailing behind the trees with its companion.

Dropping my eyes, I see a third eagle coming our way. This is a young golden, with lots of white in the wings and tail, and it apparently saw the pigeon while we were popping for the bald. It drops even lower, right down among the trees, and through my binoculars I see its wingtips lashing among the twigs as it drives onward, digging into the air and scattering a trail of leaves behind it. The eagle displays incredible agility, weaving among the saplings like an accipiter. It is coming straight at us into the clearing, no more than waist-high now, its head and shoulders glowing gold, wings set, feet down, intent on the pigeon that Mark holds motionless in the middle of the net. The world has shrunk to the width of an eagle's eye.

The eagle strikes without slowing, almost tearing the line out of Mark's hands as it grabs the lure and overshoots the net. The braided nylon pigeon line holds for a heartbeat, then breaks with a loud crack. The eagle drops the pigeon, makes a sharp right turn and glides under the nearby mist net, through a space of maybe a foot, and flaps off.

I feel dead for a moment, then bathed in frustration and disappointment. Robert slams both fists against the windowsill and curses.

Mark says nothing. The pigeon, unhurt in its leather sheathing, lies in the tall grass at the base of the lure pole where it was dropped and watches the eagle disappear to the north.

For a long time, there was concern that the number of golden eagles flying south along the Kittatinny was slowly and steadily declining. It was thought that perhaps disturbance on the breeding grounds was interfering with reproduction, or that the eagles might be picking up chemical contaminants during their sojourns south, or losing important wintering areas. But more recently, the counts have trended upward, suggesting that this small population may be recovering from the difficulties of the past. I wish I knew for sure. I do know that each year something precious passes by, like a north wind through the trees, and leaves me the richer for its passage.

Through my long autumn of hawks, I came to see the mountains as somehow separate from the valleys. This is human musing and nothing more, but the Appalachians are islands in another, very real sense. Along the summits of their highest peaks, as far south as the Smokies, exist pockets of ancient history—pieces of the ice age adrift in a postglacial world.

During the last period of climatic cooling, the ice sheets came as far south as northern Pennsylvania, gouging out lake beds in the Poconos and the Appalachian Plateau before grinding to a halt. Beyond the glacial moraines, the climate was bitingly cold and dry, and the dominant habitat was tundra; where there is now lush hardwood forest, then only sedges, stunted willows and dwarf blueberries grew, with stands of gnarled conifers in the sheltered valleys. When the glaciers melted, the forests migrated behind in their wake—first spruce, then northern hardwoods and eventually a mixture of oaks, hickories, chestnuts, white pines and, on cooler northern slopes, hemlock.

The change flowed from south to north—but importantly for the modern Appalachians, it also flowed from low to high. It is a rule of thumb in ecological science that a thousand feet in elevation is roughly equivalent to one hundred miles in latitude, so that climbing a mountain is analogous to traveling north. On the highest peaks of the Appalachians, fragments of earlier ecosystems survived like snapshots of the glacial retreat, marooned as biological islands where logic suggests they have no right to be.

In eastern West Virginia, Route 250 crosses the Cheat Mountains, a spectacularly beautiful range in the Monongahela National Forest. At the summit, a dirt road cuts north through a hardwood forest dominated by yellow birch and the carefully enunciated phrases of ovenbirds, endlessly repeating *tee-chur, tee-CHUR, tee-CHUR, TEE-CHUR!* Sunlight flickers through the new, kelly-green leaves of spring, fiddleheads uncurl, the breeze sifts among the branches and makes a sound as clear as spring water. At length the road branches, and the right turning begins to drop immediately into a narrow valley, toward a small river called Shavers Fork. The change in habitat is equally abrupt; the birches are replaced by a solid stand of red spruce, and the calls of the ovenbirds are lost among the fluting songs of blue-headed vireos and black-throated blue warblers. The sound of the breeze is heavier here, throatier, a moist breath.

The red spruce forests of the high Appalachians are natural time warps, allowing a glimpse at an earlier age in the mountains' history. Here live animals that are normally associated with Quebec or New Brunswick instead of the mid-Atlantic states. The spruce belt is famous among bird-watchers, who seek there such boreal species as ravens, red crossbills, purple finches, hermit and Swainson's thrushes and seven or eight nominally northern warblers. Snowshoe hares leave their flat, buttonish droppings on mountains as far south as Virginia (and were once found as far south as Tennessee and North Carolina), even though they are absent from lowlands below the Adirondacks. Water shrews, rock voles, an endangered race of northern flying squirrel and more than a dozen other mammals have the same southward extension in the mountains; in fact, a field guide to virtually any group of animals or plants will show a number of range maps with a thin, curved tongue arcing southwest, tracing the Appalachians.

The red spruce itself is a tree with a melancholy past (and a troubled present, but more about that later). At the dawn of the Colonial period, it was the preeminent mountain tree in the Appalachians, crowning the highest peaks from New England to North Carolina in association with balsam fir, another boreal conifer. The stands of virgin spruce were extensive; by some estimates, West Virginia alone had nearly 1.5 million acres.

While the white pine forests of the East remained, no one paid much attention to the red spruce, but by the waning years of the nineteenth century, the pineries had been virtually eliminated (so much so that Pennsylvania went from a state covered primarily with pine forest to one almost completely dominated by second-growth oak). The timbermen, looking for another source of quality softwoods, turned their eyes to the red spruce.

What followed was a relatively brief, orgiastic period of clear-cutting that smashed the virgin spruce-fir forests almost into oblivion. Often, fire finished what the lumbermen had started. On the high Appalachian Plateau of eastern West Virginia, where the Allegheny Front rears up four thousand feet above the headwaters of the Potomac River, red spruces of enormous size once formed a solid canopy. The timber was felled during the first two decades of this century, exposing the thick layer of humus and peat below, which had been laid down over the preceding thousands of years. Dried by the sun, the soil turned to tinder, and when the inevitable lightning fires kindled the slash piles, the very ground burned away to mineral soil and boulders. The result is the weird, otherworldly place known as Dolly Sods where, almost a century later the only ground cover in many places is a low mat of laurel, and the spruces are relegated to stunted clumps scattered over the tundra-like landscape. It is ten thousand acres of federal wilderness area, which ironically bears virtually no resemblance to its original wilderness condition.

You can get a glimpse—just a glimpse—of what these virgin mountaintop forests were like in the Cheat Mountains, about forty miles southwest of Dolly Sods. Near Gaudineer Knob, just across the Shavers Fork Valley, is a 140-acre tract of partially virgin timber—a pittance when compared to what once was, but seekers after lost glories must be grateful even for crumbs. At least now the trees are safe, for the grove has been officially consecrated as a natural landmark within Monongahela National Forest.

The stand is not pure but is mixed with plenty of hardwoods, and even many of the spruces are second-growth. But there are some monarchs here, too—three-hundred-year-old red spruces with circumferences of ten feet or more, climbing one hundred feet into the air. As I crane up at them, they look smooth-skinned, even though the bark is roughly scaled when seen up close. The silver, long-dead branches that cling to the lower trunks droop down arthritically, and needled branches occupy only the upper third or fourth of the trees, giving them a curiously bottom-heavy appearance. From the skeleton of a blown-down spruce, a Canada warbler voices its complicated series of twitters, but up here, so far from the road, there is no other sound save the wind, heady with the spicy, citric smell of spruce oil.

Resting on the trunk of the fallen spruce, I scuff absently at the soil at my feet, scraping a dark brown tear in the litter. Something small and wet skitters out from beneath a rotting birch leaf and vanishes, and I bend low to poke more purposefully in pursuit. Under a piece of old bark, I find it again: a tiny salamander.

The family of salamanders known as plethodons perfectly mirrors the effects of mountaintop "islands," neatly showing how isolation is a force for the evolution of new species. Plethodons, or lungless salamanders, are small, relentlessly inconspicuous creatures of the forest floor, just a few inches long, living beneath rotting logs and loose stones in the damp layer of decay where leaves become soil and minute invertebrates abound. They do especially well in cool mountain climates, where rainfall is abundant and temperatures stay moderate even in midsummer. Thanks to this moist environment, the plethodons have managed to do without lungs, absorbing oxygen directly through their permeable skin and capillary-lined throat. Given their circumstances, it is a system that works admirably.

The need for a cool, damp climate, however, may limit a plethodon's mobility, especially if it lacks tolerance for drier, warmer conditions, as many species do. While it can range widely along the ridges and knobs, the salamander may find the slightly drier oak forests of the valleys out of reach. Thus, as eastern North America heated up ever so slowly at the end of the last glacial maximum, salamanders that were ecologically addicted to the northern forests found themselves following their habitat up the slopes of the Appalachians, stranded in dozens of pockets with no way out.

Biologists since Darwin have known that isolation—physical, behavioral or otherwise—is a prerequisite for evolutionary change, and the plethodons of the Appalachians certainly were isolated. What may have been only a handful of different species to start with quickly blossomed into dozens, as populations on each mountain range (and even some individual stream drainages) diverged from their ancestral stock and took on unique characteristics.

One of those ancestral plethodons may well have been something like the redback salamander, still the most common woodland species across the East, capable of surviving in almost any forested setting as well as damp meadows and vacant city lots. The redback salamander hits four inches at most, with a flattened head, tiny eyes and ineffectual legs that look like an afterthought. The base color is gray-brown, with a long stripe of chestnut running from the head down the back and almost to the tip of the long tail; another color form, the "leadback" salamander, lacks the rusty stripe.

If the redback represents the theme, the Appalachians have produced a multitude of variations. The Peaks of Otter salamander is, as the name implies, found only on two mountains in that particularly scenic portion of the central Blue Ridge; it looks very much like a redback but has a brownish back spangled with gold. Weller's salamander, restricted to the hill country near Mount Rogers, where Virginia, North Carolina and Tennessee butt heads, takes the metallic motif a step further, with heavy squiggles of gold or silver worming down its back against a black background.

And so it goes. The Shenandoah salamander, so similar to the redback that its darker belly is the only good field mark, is found on just three neighboring mountains in Shenandoah National Park—Hawksbill, the Pinnacles and Stony Man. The Cow Knob salamander, restricted to a narrow strip of mountaintop twenty-four miles wide and just a mile across, wasn't even recognized as a distinct species until a few years ago.

There are so many endemic salamanders in the central and southern Appalachians, plethodons with tiny ranges and similar characteristics, that scientists have long been stumped as to exactly how to classify many of them. Physical appearance is a poor choice—one species of lungless dusky salamander perfectly mimics the colors of Jordan's salamander, an unrelated plethodon. For years the imitator was thought to be merely a color phase of yet another species, the Blue Ridge Mountain salamander, but biologists now realize it warrants full species status. Even within a single species, color can vary widely. The Allegheny mountain dusky salamander, found in the Appalachians from southern New England to Georgia, may have a light stripe of orange, yellow, gray, brown or chestnut along the back, which itself may or may not have a row of chevrons down the middle—and that's only in the northern half of its range. In the South, the pattern is even more variable. Increasingly, herpetologists are turning to genetic analysis to resolve the question of who's who.

Any living thing with a tiny home range is obviously at risk, and the Appalachian plethodons have had their share of trouble. The situation is so bad for two, the Cheat Mountain salamander and the Shenandoah salamander, that they are federally protected as threatened and endangered species, respectively.

The plight of the Cheat Mountain salamander is instructive. *Plethodon nettingi*[1] (the name honors biologist M. Graham Netting, who discovered the species in 1935) is known only from wooded sites, showing a clear preference for the red spruce/mixed northern hardwood forest—a dicey home choice, given the spruce forest's battered history in the region. What is more, it has a very narrow altitude tolerance, restricted to elevations of more than three thousand feet and usually confined to a limited band less than two hundred feet in vertical rise. Its entire range fits within the east-central highlands of West Virginia, and while it seems likely that in prelumbering, prewildfire days it was much more widespread, and was known to inhabit a mere sixty-eight spots in just four counties—Pocahontas, Randolph, Tucker and Pendleton. Worse yet, herpetologists guess that most of the locations support fewer than ten salamanders, making the risk of inbreeding, with its loss of genetic viability, all the greater. This is not a recipe for success.

In 1989, the U.S. Fish and Wildlife Service officially listed the Cheat Mountain salamander as a threatened species (that is, one likely to become endangered in the reasonable future). The government is required to produce a recovery plan for threatened and endangered species and, two years later, USFWS weighed in with such a plan for the Cheat Mountain salamander.

The agency's draft report notes that virtually all of the salamander's habitat has been cut, or worse, in the past one hundred years. "Wildfires during the early part of the 20th century were probably devastating to woodland salamanders, including *P. nettingi*," according to the report. In 1908 alone, every county in West Virginia suffered major fires, with many of them lasting as long as three months, until winter snows extinguished them.

"[W]ildfires in the spruce forest were particularly damaging because deep humus was burned to the bedrock. Salamanders could only survive such heat intensity if they could find deep, cool refugia beneath the ground. While it is speculative, these events could have extirpated some *P. nettingi* populations," the report concludes. Speculative, perhaps, but hardly a long shot.

A wildfire is a dramatic way to kill salamanders, but there are subtler evils. Two populations of Cheat Mountain plethodons are known to have been wiped out in recent years by timbering operations, while a third vanished under a deep coal mine. Other dangers include ski slopes, roads, wildlife food plots, power line corridors, infestations by the introduced gypsy moth and anything that removes the forest cover. Even footpaths can be a hazard; to a tiny salamander whose home range is usually less than a meter across, a wide, clear trail is an insurmountable barrier, and a few paths can separate the already fragmented populations into smaller and smaller pieces, reducing the already limited mix of genetic material even more.

The recovery plan for the Cheat Mountain salamander does not suggest such heroic measures as captive propagation and reintroduction, as is often the case with rare birds or mammals. Instead, it calls for a thorough search of the region in the likely event that other populations have been overlooked (easy to do when the subject spends its life hiding under logs), studies of the salamander's ecological needs and protection and management of its remaining habitat. Happily for the salamander, most of its range is on public land, and the U.S. Forest Service, as a federal agency, is required to manage its holdings to protect endangered species. More intensive surveys have turned up about ten previously unknown populations, including one in a new county, Grant. That's good news, but scientists have also found that redback and mountain dusky salamanders from lower elevations, which compete with the Cheat Mountain salamander, have been pushing higher and higher into the latter's range. Climate models also suggest that changes in temperature and rainfall lie ahead—an even more serious concern for a creature whose home range spans barely a yard or two, and which breathes only through its perpetually damp skin.

Anyone who lives among mountains develops a fondness for particular peaks, and over the years I've grown particularly attached to Camel's Hump, which billows up among the Green Mountains in north-central Vermont. There are fine views of it from the west, or from the north as you sweep along I-89, an odd, double-backed shape poking skyward. *Le lion couchant,* explorer Samuel Champlain called it, seeing a great feline in the mountain's shape;[2] less-cultured backwoodsmen of the eighteenth century nicknamed it the Camel's Rump, and I imagine that at least a few of them called it something pithier.

Camel's Hump is not, at 4,083 feet, the highest of the Green Mountains; Mount Mansfield and Killington Peak eclipse it by several hundred feet. Nor does it look especially big from the highway, since most of its bulk is shielded by nearer mountains, and the summit is broad rather than abrupt. But it is the only one of the trio to escape what passes for development in this ski-crazy part of the world—aerial tramways, chairlifts, the scars of slopes cut through forests of balsam and spruce and, in Mansfield's case, a cluster of radio towers with a road snaking virtually to the top.

So Camel's Hump is the last of the Green Mountains' top peaks to preserve almost untouched that rarest of Appalachian habitats—the alpine tundra. Battered by winds that carry a chill on even the hottest summer day, the summit is a realm of enforced dwarfism, the final home for Arctic plants left stranded by the global warming trend that began thousands of years ago.

The only access to the summit is by foot, and the journey is a tutorial in the way altitude and climate shape the life of a forest. A dirt road leads along the Winooski River down from Waterbury, the tea-colored water flowing beneath maples and elms. Then the track cuts into the hills, climbing along Ridely Brook through a classic northern hardwood forest of beech and sugar maple and a viridian ground fog of ferns.

The trailhead at Couching Lion Farm is a full twenty-six hundred feet below the peak, and the hike up takes about three hours. A rose-breasted grosbeak warbles as I sling on my pack and stretch my leg muscles before I hit the path, which snakes between fine old beeches, their trunks disfigured by generations of pen-knife vandals. The beech and maple zone is quickly left behind, however, and birches become the dominant species in the mixed forest. Many are yellow birches, their brassy skins as shiny as oil, while others are white birches, which everyone recognizes as the signature trees of the northern woods.

The white birch is the perennial favorite, so much so that one cannot picture New England without it, but I've always been drawn to the yellow birch. It is a more gracious tree, I think, shedding fine ribbons of bark that spiral like a blond child's curls. I love the deep, shimmering gold of the yellow birch trunks, and their habit of growing along mossy rivulets in the forest. Once, among the snarled, exposed roots of a tree undercut by a tiny stream, I found the deep nest of a Louisiana waterthrush, sealing the yellow birch's magic for me.

Wherever the trail passes through a pool of light, I find fine clouds of black-and-white moths, skittering by the dozens from sun to shadow and back again; they are spear-marked blacks, a common, day-flying species. In one passage where the birches have opened to allow dappled light onto the ground, there must be hundreds of the moths, with several big, yellow tiger swallowtail butterflies drifting casually among them. The sight resembles the jungle trails I've so often seen in the tropics, where constellations of small butterflies fill the air and a few of the big, dazzling blue morphos patrol ceaselessly, like royalty among commoners.

Above twenty-nine hundred feet the birches give way to balsam fir and red spruce, those constant companions in the northern and central Appalachians. Right at the edge of the spruce zone, where the deep green of the conifers blends with the airy green of the birch canopy, a black-throated blue warbler sings from a birch: *I-am-so-la-zeee*. Lazy or not, this is as far as it'll go up the mountain, for its species is tied to the hardwoods. Not far away, a magnolia warbler pivots on a spruce branch and eyes me fearlessly from a range of a few yards. Speared by a shaft of sunlight, the bird is dazzlingly beautiful— blazing yellow throat and breast, a black chest band, black face, white eyebrow and gray cap. As it moves from branch to branch searching for insects, it keeps fanning its black–and-whitetail and waving its yellow rump patch.

Wood warblers are among the most habitat-specific of North American birds, and the magnolia warbler prefers stands of young conifers for nesting; just as the black-throated blue will come no higher than this point, so the magnolia is unlikely to drop any lower. Nor will the blackpoll warbler, which shares the high-altitude fir/spruce forests of Camel's Hump—but where the magnolia usually nests near the top of a small tree, among the bushy cap, the blackpoll builds somewhat lower and tight against the trunk. The yellow-rumped warbler, yet another common breeding species in the conifer zone, tends to pick the outer portions of branches in older trees, fifteen or twenty feet off the ground.

The reason for such specialization seems clear—it permits the greatest use of the resource (in this case, nesting sites) with the least competition. Scientists who have studied warbler feeding behavior have discovered the same resource partitioning when it comes to where and how each species feeds. In the same conifer, one species may forage over the outer surface of the tree, hovering to pluck insects, while another will stay close to the trunk. Some tend to forage along a vertical axis, others horizontally. One species may haunt the top of the spruce, another the lower third. The differences may seem small, but they allow the warblers to find food that others, with different foraging habits, have missed. And in a broader sense, specialization of this sort is thought to promote a diversity of species by narrowly defining ecological niches.

The bunchberry along the trail is blooming, the four-bract, white flowers that betray its relation to the dogwood; I pass goldthread, false hellebore, blooming clintonia and mayflower in late bud. The trail steepens, crossing slabs of schist; the gleaming, sparkling rock looks ropy, almost like taffy that was frozen to stone. I notice that the trees are decreasing in size, and rather suddenly I find myself in a dwarf forest. Known by the German name *krummholz*, or "crooked wood," such forests are common just below timberline on the highest northern Appalachian peaks, where the ferocious winter wind flattens any tree brazen enough to raise its head above the sheltering snow. The result is a densely packed forest no higher than my waist, composed of trees that might be more than a century old.

The wind is blowing a chill out in the open, but by the standards of the Green Mountain peaks, it is a balmy day. It is also hazy; there is only a rumor of Lake Champlain to the west, and the Adirondacks beyond it are hidden completely. Today is Saturday, and there are about twenty other hikers on the summit, including a large group of teenagers who are doing a reasonable job of obeying the many "Keep Off" signs warning them to stay on the trails.

The signs are crucial, for the alpine tundra is an achingly fragile habitat, where a misplaced foot can cause irreparable harm; it is ironic but true that plants that can withstand the worst of winter cannot survive the compaction of soil around their roots caused by a few errant steps. This is particularly true of Bigelow's sedge, an unpretentious plant that looks very much like grass and grows in room-sized "lawns" in a number of slightly sheltered spots on Camel's Hump. Just the place for a weary hiker to stretch out—except that doing so may damage or kill the sedges.

If you hop carefully from bare rock face to bare rock face, it is possible to explore the tundra zone without leaving a trail of destruction. You have to keep your eyes down, however, because none of the plants, as unusual and frequently beautiful as they may be, are terribly large.

Alpine species are masters at exploiting microclimates, the fractional differences in temperature, humidity and wind velocity created by, say, an inch-deep crevice in the rocks. In order to thrive, a plant cannot outgrow its microclimate, which is one reason that most alpine species grow so low to the ground. Some are true dwarf species, while others grow low more out of necessity than genetics, like the birch tree I find wedged in a cleft. Sculpted by the inexorable wind, it has grown out and down, a green cascade cloaking the rock for six or seven feet.

These high peaks are a land of berries. Within a fairly small area, I find black crowberry, alpine blueberry, alpine bilberry and mountain cranberry, this last still holding onto some of its tiny, purple-red berries from last season. They are blooming or ready to do so, like the pink flowers of the blueberries, translucent Japanese lanterns made by Lilliputians. All of these species are low, sprawling plants, most with thick, waxy leaves designed to retard water loss.

In fact, drying is a greater threat to alpine plants than cold, and each species has its own strategy for retaining precious moisture. Labrador tea, a member of the heath family (a group well-represented on northern summits), has thick leaves that are rolled down on the edges and covered underneath with a thin coating of rusty fuzz; both features trap a space of dead air under the leaf, reducing evaporation. Lapland rosebay, another alpine heath, has leaves that are heavily scaled on the underside—although most people who see its exquisite purple, miniature-rhododendron flowers pay little heed to the bottoms of its leaves.

There are wildflowers in abundance up here, but you have to look carefully for them, because they, like everything else, survive by keeping a low profile. Most are mat-forming plants, like *Diapensia*, which grows in compact mounds of tightly packed, mosslike leaves, studded in midsummer with small, white blossoms. The similar mountain sandwort is one of the most common flowers of the alpine zone, its white petals cleft at the tips and nodding on their short stalks in the perpetual wind.[3]

Crowds bother me on mountains, so I carefully hopscotch my way over the side, into a sheltered lee looking northeast, out of earshot of the youth group that's whooping it up on top. I scan the sky carefully, for the peregrine falcons nesting to the north around Mount Mansfield and Smuggler's Notch are sometimes seen here, but at midday even the ravens are gone. Nearby, however, dark, woolly bees are clustering on the tops of several spruces, massing by the dozens. They are big, lumbering fliers, and although I don't recognize the species, I assume that their size, hairiness and dark coloration are all adaptations to the constant cold.

Alpine tundra peaks in the northern Appalachians are famous among entomologists for their endemic butterflies. Mount Washington in New Hampshire is home to a subspecies of the Melissa arctic, a small butterfly that is plain brown above but whose underwings are intricately mottled with brown and white. The same alpine tundra gardens also support a race of the Titania fritillary, a smaller, purpler version of the common meadow fritillary. Farther north, Mount Katahdin is home to a subspecies of the Polixenes arctic, which thrives in the tundra tablelands around the massif.

Many of the boreal, subarctic and Arctic butterflies in North America are also found across northern Asia, and Europe, and are known as holarctic species; their circumpolar distribution probably dates to the existence of the Beringia land bridge, which unified the northern hemisphere. The Melissa arctic and Titania fritillary are both holarctic, and although the Polixenes arctic is restricted to North America, it is found from eastern Labrador to Alaska.

Like many of the alpine plants, these butterflies must have shifted south during the glacial peak, then moved north and up mountain slopes, following the tundra that was following the glaciers' retreat. A few, like these, became stranded in the biological islands atop the highest Appalachians, where in the relatively few thousands of years since their isolation, they've evolved into distinguishable variants from the originals.

Disjunct populations of butterflies are common in mountain ranges, even if they have not changed sufficiently to warrant subspecies status. In the Appalachians they include the pink-edged sulphur, a delicate bog inhabitant the color of goldenrod, found in the mountains of West Virginia and Virginia, as well as one beaver meadow in northern Pennsylvania. The core of its range, however, lies much farther north, in New England and southern Canada, where its favorite food—sour-topped blueberry—grows.

I wish I could stay until dusk, when the Bicknell's thrushes sing their buzzy, rolling songs, which slide up and down the scale like a rock bouncing down a mountainside. But the sun has been heating the valley air all day,

Dwarfed by wind and winter cold, the forest of stunted spruce known as the krummholz *lies just below the alpine tundra zone on the highest peaks of the northern Appalachians.*

and thunderheads are bunching up over Lake Champlain and shambling my way. A treeless summit is no place for people when the lightning dances.

The alpine gardens of Vermont, New Hampshire and Maine are not the only place to find Arctic species in the Appalachians, of course, only the most southerly. Large areas of alpine tundra cover the Shickshocks in Quebec and the Long Range Mountains of Newfoundland. Here, caribou lend an even more northern flavor to the scene, and in the Long Range, big Arctic hares and rock ptarmigan are also found, the most southerly extension of their otherwise Arctic ranges.

The fate of the Appalachian alpine tundra is in doubt for several reasons. One is simple, thoughtless abuse; on easily accessible peaks like Mount Mansfield in Vermont and Mount Washington in New Hampshire (both of which have toll roads to the top), the alpine gardens have been trampled by crowds of people. Even some of the mountains accessible only by foot have taken a hard beating, like Vermont's Mount Abraham, where the tiny patch of Bigelow's sedge has been ground under heel almost to oblivion.

But there is a more insidious threat. Even under the amount of climate change we've already locked in place with the greenhouse gases now in the atmosphere, the effect on plant communities—especially those at high latitudes and high elevations—will be tremendous. What is coming if we don't change will be profound: a virtual loss of Arctic tundra (replaced by boreal forest) and an expansion of the deciduous zone up through much of eastern Canada, for example.

The Appalachian habitats and species at most immediate risk are those in the alpine zone; life zones found lower on the mountains can presumably climb upslope, but those already on the peak would have nowhere to go. True, these species have survived warming periods in the past, including the hypsithermal about seven thousand years ago, but there is no evidence those changes struck with anything like the speed we're seeing in its modern, man-made successor. Alpine plants and animals may simply not have time to adapt.

The change is already evident. By comparing what's there now with historical photographs of New England's peaks dating as far back as the 1870s, it's clear that tree line has been steadily creeping up the mountains, probably the result of not only milder temperatures, but increasing snow cover. The alpine life raft is shrinking by the decade.

The first hint of thunder reaches my eyrie, and I turn for the trail down, but not before my eye catches a patch of white. A clump of mountain sandwort has raised a dozen short-stemmed flowers to the sun, as it has since this land melted free of the ice. It can climb no farther, and its future is as hazy as the midday sky above it.

Yellow Trillium and Fringed Phacelia

Chapter 3

THE WOODED SEA

In the spring of 1776, as the Continental Congress grappled with that word *independence* in Pennsylvania, Philadelphia botanist William Bartram was wandering through the homeland of the Cherokee in the southern Appalachians. Bartram, then thirty-seven years old, had been traveling through the Southeast for three years, collecting new plants, and had finally turned away from the coast for the mountains.

His route led up the Savannah River through South Carolina, then up the Little Tennessee River ("Tenase" in his writings) through eastern Georgia and into North Carolina. Bartram stopped at Indian settlements along the way, passing through miles of carefully tended cornfields and meadows, the latter "so profusely productive of flowers and fragrant strawberries, their rich juice dying my horses feet and ancles."

At the Cherokee town of Cowe, he found "one of the most charming natural mountaneous landscapes perhaps anywhere to be seen; ridges of hills rising grand and sublimely one above and beyond another, some boldly and majestically advancing into the verdant plain, their feet bathed with the silver flood of the Tenase, whilst others far distant, veiled in blue mists, sublimely mounting aloft with yet greater majesty lift up their pompous crests, and overlook vast regions." That's still a pretty good description of North Carolina's western toe, a place of steep hills and hidden valleys known as coves, where the lesser mountains cascade out from the Great Smokies.

Without a guide, and with a good deal of trepidation, Bartram headed into those hills, following the course of a river the Cherokee called Nantahala, or "noon day sun," for its deep gorge is lit directly only at midday. Along the Nantahala he saw "at some distance, a company of Indians, all well mounted on horse-back," and recognized Atakullakulla, the "emperor or grand chief of the Cherokee."

Bartram couldn't be certain of his welcome, for the Cherokee had recently been engaged in bloody conflicts with white colonists. "[A]s they came up I turned off from the path to make way, in token of respect, which compliment was accepted, and gratefully and thankfully returned; for his highness with a gracious and cheerful smile came up to me, and clapping his hand on his breast, offered it to me, saying, I am Ata-cul-culla, and heartily shook hands with me."

Quite by accident on another spring day more than 215 years later, I came across a metal historical plaque along the Nantahala commemorating the event, not far from where the meeting took place. I knew Bartram had passed through this region, but I had not known his exact route, and despite the gravel trucks thundering past on the two-lane road, I tried to conjure up the scene. The river, famous among kayakers and rafters, looked untouched—a white-water delight ripped to porcelain foam by the rapids and lined with tall tuliptrees and maples. Then another gravel truck careened by, whipping up dust, and the illusion was gone.

But if you continue up the Nantahala for a few more miles, turn northeast across the edge of the Snowbird Mountains and drive beyond the town of Robbinsville, a latter-day traveler can still experience a little of what Bartram saw.

To the west, the Unicoi Mountains straddle the North Carolina–Tennessee border, Stratton Bald nudging up more than five thousand feet. These southern ranges are, Bartram wrote, "a sublimely awful scene of power and magnificence, a world of mountains piled on mountains." It was the middle of spring when I was there, but a cold wind was blowing and there was a dusting of snow on the highest peaks, the kind of unseasonable snap that the old-timers call the blackberry winter, for it often hits when the bramble canes are in flower.

Here, up the drainage of Little Santeetlah Creek, is the Joyce Kilmer Memorial Forest, some thirty-eight hundred acres of old-growth timber. Part of Nantahala National Forest, it contains some of the finest examples of virgin hardwood forest in the East—trees that were tall in Atakullakulla's day.

As with most remaining virgin hardwoods, the Kilmer tract survived almost by accident. Much of the surrounding woodland, on the Slickrock and Citico drainages, was clear-cut in the first two decades of the twentieth century, but the timber companies doing the cutting went bankrupt before they could chew into the Kilmer. Then, in 1925, a huge wildfire fueled by timber slash scorched a wide area but missed this tract. So the virgin forest on the

Little Santeetlah was never cut and, with the exception of a single home, was never settled. In 1936 it came into federal hands, miraculously intact.

Inside the woods the wind, which was hat-snatchingly fierce in the valleys, was quieted to a whisper by the enormous eastern hemlocks, among the finest I'd ever seen. Donald Culross Peattie, in his classic work on eastern trees, wrote that a hemlock in the wind "whistles softly to itself. It raises its long, limber boughs and lets them drop again with a sigh, not sorrowful, but letting tranquility fall upon us." In this tranquil hush I could again hear birdsong, the whistled phrases of a blue-headed vireo coming clear from above, although the understory of thick rhododendron made it impossible to see.[1]

I followed the vireo's song out the trail, along a steep hill and across a rivulet lined with purple trillium, into an opening where several big trees had fallen. Here, for the first time, I got the cathedral sense that comes with ancient trees, and all the inevitable comparisons rushed to mind—the hemlocks like columns, the roof of greenery shading out the sky. But when I sat and looked more critically (though no less raptly), I realized that the immensity of the trees, their trunks rising limbless to join the upper and lower canopies of the forest, somehow magnified the space they defined, enlarging the opening in the forest rather than restricting it. Soft light filtered through, the kind of cool illumination you get only from hemlock groves and moonglow.

Not much farther along I came across a big, moss-caked hemlock that had fallen across the trail, years ago by the looks of it. A section had been cut out so the trail could pass, and on a whim I began counting growth rings.

The first century was easy, the point of my pocket knife flicking from one ring to the next, but the farther out I got, the thinner and more compact the lines became, as the tree's growth slowed with maturity. By the end I was squinting myopically, trying to distinguish one ring from the next. Worse, the outer shell of the tree had decayed badly, and eventually the lines vanished into red-brown humus, leaving my count at 314. At a guess, perhaps another twenty-five or thirty years' worth of rings had been lost to rot, plus heaven only knows how many unrecorded years after its death before the tree fell; I know smaller, less sturdy hemlocks that have stood dead for thirty years, defying gravity. So perhaps 375 years had passed since this tree was a sapling—virtually the entire span of European settlement in this land.

Still farther, the trail reached open hardwoods, where some of the biggest tuliptrees in the southern Appalachians grow. Tuliptrees are noble at any stage of their lives, always ramrod straight and carrying a canopy of light, flickering green, but these were behemoths, lancing upward for perhaps a hundred feet, then branching into a flat, oddly graceless canopy that seemed too small for the tree's great mass. No matter; they were mighty and wonderful to see. The bark was heavy and gray, deeply furrowed vertically and ringed with close-packed bands of sapsucker holes—the work of untold generations of birds drilling for tuliptree sap.

Among the tuliptrees was another old hemlock, split halfway up into equal trunks framing the sky in a symmetrical *V*; on a boulder nearby was a brass memorial to Joyce Kilmer, whose name the tract now carries. Kilmer

The old-growth monarchs along Little Santeetlah Creek give a hint of what Appalachian hardwood forests looked like in the prelumbering boom days.

was born in New Jersey in 1886, and after a short stint teaching high school, he turned to writing. He eventually joined the staff of the *New York Times*, but his career was cut short by World War I, and in 1918, at the age of thirty-one, he was killed in action in France.

Kilmer is best known for his poetry—or rather for one poem, "Trees," the first two lines of which almost everyone knows ("I think that I shall never see/ A poem lovely as a tree ..."). In 1936, the same year the U.S. Forest Service acquired the virgin timber along Little Santeetlah Creek, these woods were dedicated to Kilmer's memory on the strength of that poem.

It is probably heresy in this corner of the world, but I have never cared much for "Trees." Probably my appreciation has been dulled by endless parodies—for surely this is the most-parodied poem in American history—but I find it drippy, rather treacly stuff. Just to be fair, however, I sat between the roots of a big tree and tried reading it again, with as fresh a mind as I could muster:

> *I think that I shall never see*
> *A poem lovely as a tree.*
> *A tree whose hungry mouth is prest*
> *Against the earth's sweet flowing breast.*
>
> *A tree that looks at God all day*
> *And lifts her leafy arms to pray.*
> *A tree that may in summer wear*
> *A nest of robins in her hair.*
>
> *Upon whose bosom snow has lain;*
> *Who intimately lives with rain.*
> *Poems are made by fools like me,*
> *But only God can make a tree.*

It was no good; here among the giants the words were, if anything, even more jarring than before. Kilmer's Victorian images of gentle, leafy maidens are very much at odds with the sense of vigor and immense power these massive old tuliptrees and hemlocks convey, more compelling than I'd ever experienced in the Appalachians. Words, for me, were simply blotted out.

The Kilmer trees are a fragment of the immense woodland that once covered the Appalachians and beyond, a forest sea that stretched from the Plains to the Atlantic and from Arctic timberline to the Gulf of Mexico. This was a woodland of many parts—boreal conifers in the north and the high mountains, northern hardwoods, oak-chestnut-hickory ridges, white pine and hemlock gorges, to name a very few—and the sum was one of the most remarkable temperate forests in the world.

Each community of trees, with its attendant plants and animals, was unique, but one natural assemblage stands out as the finest example of Appalachian forest. In the narrow valleys of the southern ranges, where the soil is rich, the climate mild and rainfall abundant, plant and animal diversity explode with breathtaking variety. These are the cove forests.

Here, an observant naturalist may find as many as fifty to one hundred species of trees, and well in excess of a thousand varieties of herbaceous plants. Not surprisingly, this diversity carries over into animals, as well—songbirds, insects, salamanders and much more. In fact, it is the most diverse temperate ecosystem outside of China.

Most forest communities have "indicator species," characteristic organisms (usually plants) that serve as field marks for the ecosystem as a whole. This is less true with the cove forest, however; while a few trees, such as Carolina silverbell, are more likely to be found in this habitat than elsewhere, there is no single dominant species. Rather, one finds a seemingly endless mixture of trees—yellow buckeyes and tuliptrees, white basswood and redbud, Fraser magnolia and red maple, a roll call of eastern hardwoods.

It seems as though the cove forest has borrowed the gentry from most of the other Appalachian forest communities: yellow birch, American beech and sugar maple from the northern hardwood forest; bitternut hickory and chestnut oak from the oak-hickory forest; eastern hemlock from the conifers. The largest species, nurtured on the fertile cove soil, stretch high, forming a closed canopy that in virgin coves may be two hundred feet above the ground. Beneath the canopy is a distinct understory layer of dogwood, redbud, sassafras, magnolia and other smaller, flowering trees; beneath that, a shrub layer of rhododendrons and azaleas that blooms even more gloriously; and finally at ground level, especially in April and May, a flamboyance of wildflowers.

This layering effect may account, at least in part, for the diversity of animal life in the cove forest. The more structurally complex a habitat, the more niches there are for animal species to occupy—and the cove forest is among the most structurally complex of the eastern woodlands. This allows for a much greater number of species (compared to a structurally simple habitat like a field) while minimizing competition.

There are countless coves scattered across the southern Appalachians, from the Virginia Blue Ridge south to Georgia, and species diversity generally increases the farther south you go, reaching its peak in and around the Great Smokies. This is the place to get to know the cove forest.

One of many varieties of salamanders in the southern Appalachians, the Blue Ridge spring salamander is restricted to the Tennessee–North Carolina border.

Great Smoky Mountains National Park can be too popular for its own good, however; the crush of visitors can be overwhelming, even in the off-season. As much to avoid weekend crowds as anything else, I ducked off the main cross-park road and pulled into a small, unmarked trailhead on the Tennessee side of the mountains. What caught my eye was a white carpet of fringed phacelia, a wildflower of the waterleaf family that is found only in the southern Appalachians. Each plant was about four inches tall, topped with two or three white flowers, delicately fringed around the edges of the petals. It was difficult, however, to concentrate on the merits of an individual phacelia, for the species grew here by the tens of thousands—a green, shin-high flood of foliage, capped with frothy white. The effect was of whitecaps lapping around the huge boulders that filled the narrow valley.

Among the phacelia were so many other species of spring wildflowers that I soon lost count as I worked my way uphill: trillium of several species, including the unusual yellow trillium with its upright, candle-flame petals and citrus scent; squirrel corn in feathery masses where the tiny stream moistened the ground; toothwort and violets and perhaps two dozen other species, blooming or about to bloom. The yellow buckeye buds were just opening, as were the tuliptrees'—the big reason that spring is the season of flowers in hardwood forests. The canopy is still largely bare; only at this time of year is sunlight abundant on the forest floor.

The stream valley was flat for a distance, then began to rise quickly; here and there big hemlocks leaned hard across the narrow gorge, and the small creek, charged with rain that fell the night before, vaulted over mossy boulders in a series of short falls. Finally, a quarter mile from the road, the valley dead-ended in a sheer, concave cliff, over which the night's runoff cascaded in a sheet. Much of the water blew to mist before it hit the ground a hundred feet down, and the base of the cliff was bathed in a constant fog.

Carefully, I began lifting flat slabs of rock and shifting sections of rotting logs. Beneath almost every one I found salamanders, often five or six beneath a single, platter-sized rock. The southern Appalachians are famous for the variety of their salamander fauna; in fact, salamanders defy the usual diversity curve seen in other animal classes, in which the number of species increases the nearer one gets to the equator. The rule holds for insects, spiders, frogs, snakes, birds and bats—but not for these small, wet-skinned creatures. The southern Appalachians hold thirty-four species, the largest assemblage anywhere in the world.

One of the first species I found was a slim, brownish salamander with orange splashed on each cheek. Ordinarily such a bold field mark would make identification simple, but the salamanders of the Appalachians are a complex lot, with great variety of color within species and clever mimicry by other species. My first, excited thought was that I'd found a red-cheeked salamander, a color variation of the widespread Jordan's salamander found only in the Smokies. But the habitat was wrong; redcheeks are normally found higher, in the fir-spruce zone, and when I looked more closely, I saw that the salamander's hind legs were noticeably bigger than its forelegs, a feature not found in redcheeks. This was an imitator salamander, a member of a completely different genus than the redcheek, but which almost perfectly matches its appearance. Like the red-cheeked salamander, it is restricted to this small corner of the Appalachians.

Mimicry is common in nature and is generally assigned to one of two categories, both named for the naturalists who first described them. Batesian mimicry is what most people usually think of: a harmless or edible species mimics a toxic or dangerous one to gain protection. Examples abound, from palatable viceroy butterflies mimicking toxic monarchs to nonpoisonous milk snakes mimicking venomous coral snakes and hover flies that ape the patterns of bees.

Müllerian mimicry, on the other hand, seeks to explain why so many toxic or venomous species look generally alike. Bees, hornets, wasps and other stinging social insects are good examples—forms that are only casually related share the same yellow-and-black bands, a motif that advertises rather than camouflages. Warning coloration serves notice to predators: Do not eat me. Sharing the same general warning pattern makes the job of avoidance easier for predators and safer for their prey, which might otherwise perish while teaching the lesson. As if all this weren't confusing enough, "mimicry complexes" arise, in which a crowd of species—some edible Batesian mimics, some toxic or venomous Müllerian mimics—all use the same general pattern.

So are the Jordan's and imitator salamanders examples of Batesian mimicry or Müllerian? On the surface the relationship is Batesian, because the Jordan's is toxic while the imitator is not. But the answer may go deeper than that; there are other southern Appalachian salamanders with at least superficially similar patterns, and no one will be surprised if they are also enmeshed in a complex of warning, camouflage and deceit that we've barely begun to unravel.

Within the grand tapestry of the Appalachian deciduous forests there are three main themes. In the south are the cove forests, in New England the so-called northern hardwood forest of maple, beech and birch. But in between is the most widespread of the three, and perhaps the most characteristic of the range's broadleafed communities: the Appalachian oak forest.

From southern New York to North Carolina, the oak forest flows over the middle hills, those too high and dry for cove woods but too warm and low to support northern species. It is often a hyphenated woodland, the oak-hickory forest, for shagbark, bitternut, pignut and mockernut hickories thrive in the same warm, dry conditions that oaks prefer. At one time, the more northern areas were oak-chestnut forest, until an introduced blight killed the great chestnuts, reducing them to relics that sprout from persistent rootstock but never grow bigger than saplings before they are killed back again.

This is the forest that I know best, and even when I'm far from home, I have only to close my eyes and conjure up the back-of-the-throat tang of old, fallen leaves, the ringing call of a pileated woodpecker and the rippled hills with their coverlet of oaks.

There is no better time than autumn to appreciate the mix in an Appalachian hardwood forest; the trees break ranks with the uniform greens of summer, taking on their own characteristic tones on their own unique timetables. Someone who has grown up in the hills can look at a distant mountainside and tell from a glance what grows there—the yellow mounds of hickories, the taller, more open canopies of tuliptrees shining a fainter shade of gold (and, on drier ridges, usually tracing the courses of streams and spring seeps), the deep crimsons of red maples, the maroons of ashes, the rich bronzes and rusts of oaks and so on.

I took a hike on the ridge near my home yesterday, on an October morning when the spiderwebs across the trail hung thick and ropy with dew, and the grouse boomed up from blackberry thickets already glowing purple with the changing color. The woods held the liquid, stained-glass quality of light that comes only on one or two days each autumn, when the sky is scrubbed clear of any haze, the foliage has reached its most luminous stage, and sunlight bounces off a million tiny, brilliant lenses.

But surrounded as I was by such beauty, what held my attention was the ground. Walking through the woods was like hiking on marbles, so thickly had the oaks broadcast their acorns this year. There were the small, round acorns of bear oaks, many of the large, goblet-shaped acorns of red oaks and the oblong nuts of chestnut and white oaks. I forgot about hiking and did some crude science. Each step of my size eleven boots, on average, covered 16 acorns; extrapolating from that, I estimated an astonishing density of 432 acorns per square yard or nearly 2.1 million per acre. At about 5.5 grams a nut, that's roughly 12 tons of acorns, and while such quantities were not uniform through the forest, even the sparser patches had plenty.

The acorns do not fall this lushly every autumn; this is the heaviest crop I can remember in twenty years, and last year's was paltry by anyone's standards. Nor are acorn crops consistent from place to place. While the oaks along the ridges of eastern Pennsylvania were bearing extravagantly, many areas west of the Susquehanna and north into the Appalachian Plateau region were having a decidedly meager year.

This boom-and-bust mix is not accidental but is part of the trees' genetic strategy for making the most of their reproductive potential. The phenomenon is known as "masting," and it's common in nut-bearing Appalachian trees (it occurs, as well, in such conifers as white pines and spruces). The underlying rationale is simple: If the trees bore moderate but predictable quantities of nuts each autumn, they would support higher populations of so-called seed predators such as squirrels, white-footed mice, jays, deer and turkeys, thus reducing the number of acorns that survive to sprout. This is a

The fallen leaves of black birch carpet the ground with yellow on an autumn day in the Pennsylvania hills.

matter of life and death for the trees, which expend great amounts of precious energy and nutrient reserves to produce their seeds, only to have the vast majority fail to germinate. For instance, popular belief aside, squirrels are not the diligent planters of trees everyone thinks they are. That includes Henry David Thoreau, who advocated a national holiday of thanks for their supposed efforts: "Consider what a vast work these forest planters are doing! So far as our noblest hardwood forests are concerned, the animals, especially the squirrels and jays, are our greatest and almost only benefactors." Squirrels cache nuts, certainly, but careful observation has shown that they subsequently dig up and eat virtually everything they stash away.

The trees understand the stakes better than Thoreau did. Since a predictable nut crop could encourage even greater numbers of squirrels, the trees have opted for an erratic, unpredictable fruiting pattern. For several years the nut crop will be low, occasionally almost nonexistent, keeping the population of seed predators at a similarly low level. Then comes a year like this one, when the trees explode with acorns. The abundance of food simply overwhelms the animals that feed on the nuts—and enough acorns survive to sprout and grow the following spring.

Yes, the populations of squirrels rise the next year, but since masting usually occurs in a two-to-seven-year cycle, chances are the following autumn will see a rather poor crop, and starvation will again pare down the animal population. One study of white-footed mice in Virginia showed that the mouse population went from about one per acre to forty-two per acre after a heavy acorn crop—then fell back to just five per acre when the supply dried up the next year.

There is a slow, subtle dance between any predator and its prey, one played out over generations, and so it is between acorn and squirrel. Some species of oaks have evolved high levels of tannic acid within their nuts, a protection against mammals that eat too many (tannin interferes with protein absorption). Some mast-feeding species, though, have evolved their own resistance to high tannin concentrations, squirrels among them.

Other oaks have opted for a more mechanistic approach. The white oak group (which includes white, post, chestnut and chinkapin oaks, among others) produces nuts that are low in tannin, and thus seemingly less well defended than the tannin-rich acorns of red, black and scarlet oaks. But the white oak acorns sprout soon after they're dropped, boring a deep taproot into the ground and quickly transferring the stored energy of the nut to the root, where it is out of a squirrel's reach. Red oak acorns, laced with tannin, are content to wait until spring, and thus can be cached by the squirrels for winter use.

That's only one side of the dance, however. Squirrels can circumvent the white oak's strategy by cutting out the plant embryo at the top of the acorn, killing the seed while preserving its food value. Such a decapitated acorn can be safely stored away for use later in the season.

The white oak has not yet evolved a defense against this particular attack, perhaps because it doesn't have to. Killing acorns is a learned behavior, one that seems to take a year or more to develop in gray squirrels, and since most die before their first birthday, the majority of white oak acorns sprout unhindered each fall. But if the behavior becomes genetically encoded among the squirrels, so that it is instinctive rather than learned, then the oaks will be forced to find a way around it. In such joined ways does evolution often move, an imperceptible push and tug, slow move and countermove.

A couple times each fall, I become a seed predator myself, heading out to the woods with a cloth sack to collect hickory nuts. The timing is always a little dicey, for I have to beat the squirrels to the nuts yet not arrive too early before most of them have fallen free of their thick, greenish hulls. I must also avoid being too greedy, because it's much easier to collect hickory nuts than it is to open them and pry out the nutmeats.

After an evening of carefully cracking open the pale, hard nuts of shagbark hickories, I have a deep respect for the persistence and jaw power of a squirrel. Gray squirrels, which have strong jaws and teeth, reduce a hickory shell to pieces, leaving the edges scalloped with tooth marks. Flying squirrels and chipmunks, with their weaker jaws, concentrate on a single opening, enlarging it enough to reach the oily meat inside; their work is smooth-edged and delicate.

Because I like to use the nutmeats in holiday breads, I try to extract the largest pieces possible, so I labor over each hickory nut with hammer and nutpick. Indians and the pioneers who learned from them usually took a more direct, mass-production approach to hickory nuts, as described by William Bartram during his travels among the Creek: "They pound them to pieces, and then cast them into boiling water, which, after passing through fine strainers, preserves the most part of the liquid: this they call by a name which signifies hiccory milk; it is as sweet and rich as fresh cream, and is an ingredient in most of their cookery, especially homony and corn cakes." Such "hiccory milk" was known as *pocohicora* among some tribes in Virginia, and the term doubtless gave rise to the tree's modern name.

I made pocohicora once, following the original instructions by casting all the shell bits, most with chunks of nutmeat still wedged inside, into a pot of boiling water. As promised, the meats soon detached and rose to the surface, where they could be easily skimmed off with the thick oil. But while it was easy, I didn't care for the consistency of the final product, so I stick with my tedious hand-cracking.

I've never run across anyone else on my nutting expeditions, which is an indication of how much country life has changed in the last century. Time was that whole families took to the woods each October, burlap sacks tucked under Father's belt, and everyone with a favorite "nutting stick." My great-grandfather talked often about going nutting as a child in the late nineteenth century—flinging his stick high into the trees, bringing down a cascade of hickory nuts, or walnuts, or chestnuts, then scrambling with the other children to scoop up the bounty. In his last years he still collected nuts, bringing paper grocery bags of walnuts to our house, where I was delegated the task of hulling and cracking them. This I did reluctantly, for walnut hulls were a source of dye in colonial days, and even with a pair of old work gloves for protection, the brown liquid quickly seeped through and stained my fingers the color of strong coffee.

We humans are poorly fitted for converting mast to meat, but other animals excel at the task. Wild turkeys swallow walnuts and hickory nuts whole; the nuts pass to the powerfully muscular gizzard, where they are crushed as easily as a child crumples a wad of paper. The power of a turkey's gizzard is remarkable. In one experiment, a turkey crushed metal tubes that required five hundred pounds of vise pressure to collapse; in another, an eighteenth-century Italian scholar fed surgical scalpel blades to captive turkeys. The blades were reduced to harmless steel filings in sixteen hours, with no ill effects on the turkeys.

Turkeys will scratch up enormous patches of forest floor seeking mast, their work looking like that of children armed with rakes. Deer consume great quantities of acorns as well, preferring the less tannic nuts of white and chestnut oaks over the tannin-laced acorns of red, scarlet and black oaks when they have a choice (this even though red oak nuts are higher in essential fats). Black bears are great, ebony mast vacuums in a good nut year, siphoning down acorns and beechnuts that will see them through their winter denning period.

But in a mast year like this, much of the provender of the forest will lie untouched except by insects and mold. It was not always thus.

In 1808, in Kentucky, ornithologist Alexander Wilson watched the sky blacken with birds—a flock he estimated at a mile wide that passed him for four hours. They were passenger pigeons, easily the most abundant species of bird in North America and likely the most numerous land bird in the world. Based on his observations, Wilson calculated the flock's number at 2.25 billion birds. They, and their kin, were the single greatest living expression of the bounty of the wooded sea.

"The pigeon was a biological storm," wrote Aldo Leopold in 1947. "He was the lightning that played between two opposing potentials of intolerable intensity: the fat of the land and the oxygen of the air. Yearly the feathered tempest roared up, down and across the continent, sucking up the laden fruits of forest and prairie, burning them in a traveling blast of life."

The passenger pigeon has become, with the bison and the dodo, the ultimate symbol of mankind's senseless waste. But that comes later, with the story's tragic, greedy ending. Less often is the pigeon seen for what it first was—a biological phenomenon of breathtaking sweep, unmatched anywhere else on the planet.

One of two doves native to the Appalachians, the passenger pigeon averaged about sixteen inches long and looked like a large copy of the smaller (and still abundant) mourning dove. Beyond size, the most notable physical difference was color, for the passenger pigeon had a bluish tone, especially on the rump, where the color was bright; males in particular also had a deep, wine-red wash on the breast, lacking in mourning doves.

But however similar the two species were in outward appearance, in lifestyle they were polar opposites. Mourning doves are at best loosely colonial, usually nesting alone and flocking by the dozens, or at most a few hundred. Passenger pigeons lived every moment of their lives enmeshed in dazzling numbers. No other North American bird was as capricious in its wanderings. No one knew when the great flocks would come or where they would go; their paths were as unpredictable as the wind, and their nesting grounds equally erratic. Their range stretched as far east as the Atlantic (at least during the first years of settlement) and as far west as Montana. Flowing between summer and winter, they would rise to Canada and ebb to the Gulf Coast, but their core range was the Appalachians west to the Great Lakes and northern prairies, where the forests and grasslands merged. Across this wide tapestry the flocks struck like lightning, and with as little warning.

For all their numbers, the pigeons existed within only a few mighty flocks—perhaps no more than ten or twelve, each numbering in the hundreds of millions of birds. They would gather in spring for nesting, congregating in two main arenas, one in New York or Pennsylvania, the other in the Great Lakes region; the exact location varied from year to year, depending on where the previous autumn's mast crop had been heaviest.

Little has been written about the pigeon's place in the ecosystems of the East, but its impact must have been enormous. Several billion birds feeding through a forest would reduce the mast available to other nut-eaters[2] and would seriously reduce tree regeneration; the fractured, mangled trees, collapsed under the sheer weight of the roosting or nesting birds (to say nothing of the vast quantities of guano), would have provided a flush of sunlight and fertilizer for new plant growth, benefiting grazers and browsers such as elk and deer.

And such a prey base, however erratic in its movements, must have supported a train of predators. Some ornithologists speculate that peregrine falcons in the East may once have been passenger pigeon specialists, and likewise goshawks in the more northerly breeding areas. "The howling of wolves now reached our ears," Audubon wrote after watching a mass pigeon slaughter at a roost along Kentucky's Green River, "and the foxes, lynxes, cougars, bears, raccoons, opossums, and pole-cats were seen sneaking from the spot, whilst Eagles and Hawks, of different species, accompanied by a crowd of vultures, came to supplant them, and enjoy their share of the spoil."

As early as 1634, word of the pigeons reached Europe. Captain Thomas Young, exploring the valley of the Delaware River, reported "infinite number[s] of wild pigeons," a fact William Penn reiterated in the 1680s as an enticement to settle in his new colony. Perhaps even more amazing than the migrant flocks were the pigeon "cities," mass nestings that defy belief; some were miles wide and thirty or forty miles long. Densities in some colonies were as great as a hundred nests per tree; an encampment, as these sites were sometimes known, in Wisconsin in 1871 contained an estimated 136 million birds spread out across 850 square miles of forest.

By that late date, the pigeons had been exterminated in many parts of their eastern range and were under ballooning pressure in the Great Lakes region. They were shot for home use and market, and fed to hogs; bonfires dosed with sulfur killed them on their roosts, and the pigs were sent in to clean up the leavings. The telegraph and railroad may have sealed their fate, for the flocks finally lost the element of surprise. Professional "pigeoners," who shot them for market or collected the squabs for sale as live targets at shooting matches, could now track their progress, descend on them from every quarter and easily ship off the carcasses.

"The season, commencing in April, was profitable for only a month," Peter Matthiessen wrote in *Wildlife in America*, "and by June the markets were glutted, the pigeons were scattered, and the hunters had largely departed, leaving behind a rancid wasteland of ground white with guano, of broken trees, nests, eggs, and blue-feathered, fly-blown forms too shattered to ship, of starving squabs, of maggots and silent fur-clawed and beaked prowlers."

Worse, the constant pressure drove adult pigeons from their colonies, so that in some years there was a complete nesting failure. David Blockstein and Harrison Tordoff, in their reexamination of the pigeon's extinction, point out that the great collapse occurred over just two average pigeon lifetimes, probably the result of harassment and breeding failure more than actual killing.

There has been a lot of speculation about the demise of the passenger pigeon, particularly its sudden, final free fall into extinction. As already noted, as late as the 1870s they were still nesting by the hundreds of millions in a few areas, yet by the turn of the century, they were essentially gone from the wild. A few were collected around the turn of the century, including pigeons shot in 1902 in Pennsylvania and Indiana. President Theodore Roosevelt saw a flock of a dozen in Virginia in 1907, the last reliable report of any in the wild. Some remained in captivity but bred fitfully; the last, a female named Martha, died on September 1, 1914, in the Cincinnati Zoo. It is supremely ironic that the last passenger pigeon should have had a name—this, the bird that, more than any other, immersed individuals into the whole.

The idea that passenger pigeons could disappear was difficult for many to swallow. Some contended that the birds had flown elsewhere (Australia was mentioned as a possibility). Even with the evidence of their eyes, few could credit that human action had brought the flocks to nothing. Epidemic disease has been suggested as a reason for the final disappearance, but it seems more likely that the pigeon had simply become so specialized for survival in immense numbers that reduction meant oblivion. The dwindling survivors did continue to breed, at least at low levels; juvenile birds were found right to the end. But because they were adapted for life within the insulating mass of the colony—nesting in exposed, vulnerable spots and laying only one egg a year—the remaining pigeons were unable to keep up with even the natural drain of predators and accidents, let alone humans. They were doomed.

"Like any other chain reaction, the pigeon could survive no diminution of its own furious intensity," Leopold wrote. "... Today the oaks still flaunt their burden at the sky, but the feathered lightning is no more. Worm and weevil must now perform slowly and silently the biological task that once drew thunder from the firmament."

There are many things I mourn, living in this latter, lesser day. I mourn the loss of the mighty forests, the vanished elk and bison that no longer tramp the Appalachians. I wish in vain for a wolf's song on a midwinter's night and must be content with a yapping fox. But most of all, perhaps, I grieve that I will never see pigeons. I have sat with closed eyes and tried, straining my imagination, to rebuild in my mind the multitudes that gathered the sky to themselves and blew a wind beneath their assembled wings like the roar of a storm. But it cannot be. Some things we lose but still remember; others pass away into a limbo that defies recall. The pigeon is in this farthest recess, reduced to quiet words on a page, growing fainter with the years.

The leaves started to fall in earnest this morning after a cloudy, drizzly night when they hung limp and turgid. With dawn came the wind, a bullying northwester that tore them from their weakened moorings and flung them skyward, flickering like the wings of birds across my vision and making the forest floor shimmer with movement.

All through the previous week, the leaves had deepened and intensified, as the last vestiges of green were banished and the warmest tones came singing through, golden beneath an autumn sky of gentian blue. The weather was perfect for foliage, the days warm with bright sunshine, the nights cool but not freezing. This allowed the trees to continue to manufacture sugar by day, but the nighttime chill retarded absorption of the sugar into the stems.

Trapped in the leaves and exposed to sunlight, the sugar began to break down chemically into a pigment called anthocyanin, producing the most vivid oranges and reds; further chemical reactions create the bronzes and purples. Other colors—the yellows, the rusty browns—were present all summer in the leaves but were masked by the pervasive green of chlorophyll. Each year as autumn gains momentum, though, the tree withdraws this precious chemical into its twigs and bark, pulling the drapes back on the palette hidden within each leaf.

Color also can be genetically determined; certain individuals may produce particularly intense colors year in and year out, regardless of weather, or be hereditably bland. Even soil conditions play a role in the fall display. Red maples, for example, tend to show their deepest, richest colors in moist, acidic soil.

The trees do not turn colors all at once, of course. The process is prolonged, lasting for nearly two and a half months from first blush to last leaffall, and the progression is fairly predictable, at least within the inherently fickle parameters of any natural event.

The first of any given species to change are the sick, the injured, the stressed or the dying—trees that haven't the energy to stay active into autumn. They may be growing outside their normal habitat, or have suffered through a bad insect infestation, or been partially barked by a rabbit. You can see this phenomenon along a highway in midsummer, after the maintenance crews have chopped and sprayed the undergrowth edging the road. The staghorn sumacs, mangled by brush-hog blades, sickened by exhaust fumes or sprayed with herbicide, will turn their trademark crimson months before they would have done so naturally. For a few days, in death, they become a portent of the coming season.

Likewise, unusual weather can affect the timing of the foliage change. This is particularly true of drought; deprived of water and facing winter, a season when soil water becomes progressively harder to obtain, the trees shut down early. A few years back a drought gripped the central Appalachians through a winter, spring and summer, and by the last weeks of August, the sweet birches were giving up the struggle. Their yellow was insipid, for sugar production is weakened by the lack of water, and presaged the rest of autumn—early, washed-out foliage and hills that were naked well before their time. The trees were cutting their losses, sacrificing the chance for additional energy production and food storage against the likelihood that the drought wouldn't break anytime soon. Better to settle into winter dormancy and hope for wetter days come spring.

This was a normal year, in contrast, with rain in reasoned amounts and at nicely spaced intervals. As is usually the case, the first sparks of the autumn pageant were the purple-reds of a few precocious black gums, an understory tree that grows hereabouts on the dry ridgetops beneath a canopy of chestnut, red and white oak. Although they grow nicely in wetter areas, the gums on the ridges tend to be small trees, forearm-thick and covered with checkered, pale gray bark, with high, scraggly branches that spread out at right angles to the trunk. Forming groves, they create an open tent of maroon when they change, all the more dramatic because the rest of the forest is still quite green.

The timing is not capricious, for at the same time that the gums are turning red, their fruit is ripening. Growing in clusters of two or three on the twigs, each fruit is a small oblong, perhaps three-quarters of an inch long, turning a glossy purple-black as it ripens. Technically, these are drupes, the name given to fruit consisting of a fleshy covering over a single pit, rather than multiple seeds. The drupe flesh is bitter to a human palate, but animals do not mind; gum is a favorite food plant for dozens of bird species, which relish its drupes for their high fat content. The early color change of the leaves, ecologists believe, serves as a visual signal to the birds that the fruit is ripe and ready.

This phenomenon is known as "foliar fruit flagging," and it is common in the Appalachians during autumn. Many trees and shrubs that bear high-fat fruit announce their availability by turning color (often red, a hue easily seen by birds) weeks before nut-bearing trees like oaks and hickories. They include dogwoods (both the flowering dogwood with its star bursts of red fruit and the bushier red-twigged and alternate-leafed dogwoods, which bear blue fruit) as well as sassafras, spicebush, magnolias and Virginia creeper.

The birds get a meal, but the benefit to the tree is less obvious, until one considers the workings of the avian gut. Birds process food rapidly, especial-

Sleek and well-fattened on the acorns and other mast, a white-tailed buck is ready for the grueling endurance test of the autumn rut, when he'll need every calorie of stored energy.

ly fruit, which may pass from mouth to anus in as little as forty minutes. The nutritious flesh is stripped and digested, while the seeds pass harmlessly through the stomach and intestines to be deposited at some far-flung place. For an organism rooted firmly to the ground, dispersion through an animal's digestive system is one of the most effective means available.

While resident birds like grouse and turkeys feed on the color-flagged fruit, most of it is consumed by migratory songbirds. Because the fats in the fruit provide nearly twice as much energy as carbohydrates, the fruit is an ideal food for small creatures that need copious amounts of energy but cannot afford to be weighed down by heavy meals. Nor do the trees skimp on the bribe; spicebush fruit may be 35 percent fat and a dogwood's 24 percent, representing a significant investment on the tree's part. Yet the deal works both ways, as the abundance of dogwood and the regular migration of fruit-eating songbirds show. Indeed, it is hard to imagine how some songbirds would manage their two-thousand-mile trips without a ready supply of fruit along the way.[3]

The birds display a definite preference for certain fruits, and laboratory experiments have shown that this preference is tied directly to fat content. Early in autumn, the high-fat offerings like gum drupes and dogwood fruits are taken first and most frequently—not just because they are more nutritious, but because they may be more perishable. Other species that fruit abundantly and grow cheek-to-jowl with the high-fat trees are ignored at first; these include poison ivy, American mountain ash, deciduous hollies like winterberry and many viburnums. Instead of a fat content exceeding 25 or 30 percent, most of these have less than 10 percent fat and are skipped over by the long-distance migrants.

Resident bird species, which haven't the extreme energy demands of the travelers, use these lower-quality fruits as a fallback food supply, turning to them late in winter when better sources are exhausted. And while the foliage of these trees and shrubs turns colorful—sometimes spectacularly so—it normally does so in the general crescendo of the forest, not standing apart as the fruit-flagging species did weeks earlier.

Only a few Appalachian birds eat significant amounts of fruit on a year-round basis, and only one, the cedar waxwing, qualifies as a true frugivore, reliant on fruit for most of its diet. The rest are seasonal opportunists, switching from a protein-rich menu of insects and other invertebrates in summer to fruit in the fall and winter. They must make compromises to do so, including not processing the fruit as effectively as waxwings. Robins switch from a diet of 90 percent invertebrates in spring to one of 90 percent fruit in fall, even though they lack the essential enzymes needed to break down the complex sucrose molecules in many fruits—an enzyme that waxwings possess. There are some indications that seasonal fruit-eaters like thrushes may partially compensate by growing longer guts during fall, permitting them to squeeze the greatest amount of nutrition from the food they eat.

I've often sat in a black gum grove in late September or early October and watched, in the first hours of daylight, as flocks of migrating songbirds rippled through. Not all are fruit-eaters, of course. The warblers hunt for insects while largely ignoring the blue-black drupes hanging all around them. But many of the others—thrushes, waxwings, catbirds, tanagers, orioles and rose-breasted grosbeaks among them—stop to gorge. And mixed with the whispered calls and the shuffle of wings are the regular, soft splatters of bird droppings hitting the leaves—the payoff for the trees, as another generation of seeds gets its start.

The fat of the land—not only fruit, but bugs and seeds and worms and a hundred other foods—carries these birds far, but their migration is mostly unseen and unsung. Hundreds of thousands of people flock to the mountains to watch the passing hawks or gather on the coasts to see the multitudes of ducks and geese, but songbirds pass us by like a dream, flying through the dark night with only the faintest tinkle of flight notes drifting down to hint at their presence. On a clear night in September, I'll lie back on a blanket and train my binoculars on the face of the full moon, hoping to see the tiny winged shapes flashing across the disk. I can hope for only the most fractional of glimpses, birds visible barely long enough to register as birds—or so one would think. Yet each of the handful of times it happens in the course of an evening, the image burns itself into my retina, the small black silhouettes etched against the white moon, sharp-edged and impossibly clear. It is like discovering a secret, and I catch my breath every time.

Songbirds fly at night for several reasons. The night air is generally less turbulent, so they needn't waste as much energy battling winds and contrary currents; besides, since they cannot soar like hawks, the rising thermals of midday would do them no good anyway. Too, they are less exposed to danger at night, since hawks are exclusively diurnal migrants, and the owls that also migrate at night would not be agile enough to take songbirds from midair. (At least, presumably they would not be. We really know little of these grand nighttime movements, and it is inadvisable to make too many assumptions.)

The night air is cooler, especially at high altitudes, and this is a critical concern for a migrating songbird. Because it cannot soar, a warbler or thrush must flap its wings constantly in order to stay aloft—at a rate of at least several beats per second. Exercise on this scale generates dangerous levels of body heat, which the chill night air dissipates more effectively, protecting the bird from fatal overheating. Perhaps even more importantly, in cooler air they lose substantially less water, making dehydration less of a threat.

Most songbirds that breed in the Appalachians take an overland route south through Canada and the United States, although they may not stick to the mountains; because they aren't dependent on deflection wind currents like hawks, they are free to take the most direct route or to follow the coast, as tens of millions do. Those that continue to Central or South America have a choice to make when they hit the Gulf of Mexico, however—curve west to stick with land, adding hundreds of miles to their journey, or continue generally south across more than five hundred miles of open water to landfall on the Yucatan or the eastern Mexican coast.

A surprising number opt for the over-water route (it is not, of course, really a "choice" in any individual sense, for the path is genetically imprinted in each species). This is remarkable enough, especially given such extraordinarily small migrants as ruby-throated hummingbirds. But even more surprising are birds like the blackpoll warbler, a common breeder in the northern Appalachians. Many blackpolls take an oceanic flight path, leaving the New England coast and flying nonstop down the western Atlantic, gaining altitude steadily until they are passing over the West Indies at twenty to twenty-three thousand feet, where the flocks show up on radar scopes as diffuse blobs of glowing green. Then they begin to drop, settling in on the northeast coast of South America some eighty to ninety hours later.

To say, as ornithologists have, that this is the equivalent of a human being running four-minute miles for three and a half days still shortchanges the enormity of the bird's accomplishment. It does this without rest, without refueling, without a drink of water. That even a few manage the feat is astounding, but the fact that most blackpolls take this course means that it is the preferred option. As insane as it sounds to us, this annual marathon works. That alone should leave us dumb with wonder.

Brook Trout

Chapter 4

FROM FERTILE WATERS

On an unseasonably hot May afternoon, the song of a hooded warbler lured me through the woods and into a rhododendron thicket. For no other reason than that I was bored with walking the marked trail, I plunged off the path after the bird and tried to force my way through the tangle, a tortured maze of branches and twisted trunks. No wonder the southern highlanders called this kind of place a "laurel hell."

It was like squeezing through an endless series of jail bars, tripping over hidden roots, my daypack constantly hooked by reaching branches. The second time my face was caught by a springy, rebounding limb, leaving a red welt across my jaw, I began to regret my decision.

There was nowhere to go but forward, however, so I pushed on. The rhododendrons were only thirty or forty yards deep in any case, and I could hear the warbler even louder from what looked like a small opening ahead. When I finally broke through, it was into a small bowl cupped between the high ranks of rhododendron, filled with sunlight and spicebush and the sound of trickling water. The warbler, the color of dandelions with the black cowl of a monk, fretted at me from the undergrowth.

The bowl held a spring, bubbling up from the ground, forming a tiny stream and disappearing beneath the blue-green darkness under the rhododendrons. The spring was quite small—about three feet wide on all sides—bounded by sensitive fern and sphagnum moss, with a smooth, fine-grained bottom. Where the water rose up, the sand churned and boiled, making patterns with the grace of flames. Such springs are common in the Appalachians wherever Silurian sandstone underlays the forest and has been eroded into thin beds of sand finer than salt. The constant flow of water sweeps away any small, light particles that would make the spring turbid, while heavier grains sink and stay sunk. Only those granules of just the right weight take part in the spring's endless dance—light enough to be borne upward by the rise of water, heavy enough to sink back a moment later and be recycled into the mix. Watching, I tried to find the correct analogy for the shapes the spring created. Sometimes they were like bubbles, at other times almost perfect little geysers, occasionally even like the arching, cresting flares that erupt from the skin of the sun. But mostly they looked like the unique gurgling of a sand spring.

We are fascinated by beginnings. We celebrate our births, the inaugurals of our countries, the anniversaries of our marriages. That's why, I think, we look with such delight on the headwaters of a creek. We like the certainty of starting points and ending points, serving as bookends that bracket everything we touch, perhaps because our own lives are bracketed. It is this same fascination that makes me stare at the spring of a tiny, unnamed stream, a fascination that sent men to their deaths in the search for the source of the Nile.

Hemlocks shade a headwater stream in northern Pennsylvania, keeping the summertime temperature low enough for coldwater species like brook trout and mayflies.

I pulled a rumpled topo map from my pack and found the wispy thread of blue that marks this stream's course; it snakes over the flat along the ridgetop and falls down the mountainside, joining other threads to form a creek that eventually becomes big enough to merit a name. I noticed that it cuts the corner of the trail I had been hiking and rejoins it just shy of the car, so I decided to follow it and see what secrets the rhododendrons were hiding.

For a hundred paces the creek stayed tiny, a rivulet no wider than my two hands stretched thumb-to-thumb, banked with moss and twisting between the roots of hemlocks. Then it started gathering the flows of other sand springs and seeps, the contributions of equally tiny creeks. The slope steepened, and the water flashed into small waterfalls—ankle-high, then knee-high, finally waist-high. The outswept branches of the hemlocks formed a fine chamber, enclosing the liquid sound, and the pools got larger and deeper the farther down the hill I climbed.

I kept an eye on the water, hunting among the fractured reflections and swirling foam for any movement that was organic, and eventually I saw something that my mind told me wasn't just the chop of water. I sat against a hemlock and waited, and sure enough, several minutes later a little cigar-shape moved out of the concealing churn below the falls and into the quieter water near the tail of the small pool.

I carefully raised my binoculars and turned the center wheel as far to the right as it would go. The range was so close that I was barely able to focus on the shape in the water, which in the field glasses became a six-inch brook trout, riding nose-first in the current of the stream.

I've snorkeled on barrier reefs alive with butterfly fish and wrasses the color of neon signs and dipped dazzling tropical tetras from jungle rivers, but I know of no fish as lovely as a wild brook trout. This one, probably a male by its colors, was as pretty as they come. The back was dark green, the same shade as the shadows beneath the rhododendrons, scrawled across with golden spots and swirls. On its sides the brassy spots were joined by others of scarlet, encased in halos of pale blue. The sides and flanks deepened to red, and the orange fins were the color of sugar maple leaves in October, each racing-striped with thin bands of black and white on the leading edges. The uphooked jaw had a stripe of white along the underside that I could see moving as the fish opened and closed its mouth rhythmically, flicking its fins in the current. The whole effect was that of an ornament made of hammered metal and fired enamel and miraculously given life. I know of no other fish as beautiful, and if this is regional favoritism, a subconscious wearing of my Appalachian heart on my sleeve, so be it. Everyone should have such biases, and should be forgiven them.

The brook trout is the only trout native to much of the Appalachians;[1] the rainbows and browns that are dumped each spring from hatchery trucks are western and European imports, respectively. The hatchery trucks also dump domesticated brook trout, insipid, washed-out imitations with fins and noses blunted from rubbing against concrete raceways. They are a denigration, and as far as I'm concerned, the sooner they are caught and removed the better.

Purists will demur, pointing out that brook trout are technically char, not trout, and they are right. It doesn't really matter; by the same token, rainbow trout are now classified as Pacific salmon, while the Atlantic salmon is, taxonomically, a trout. Suffice it to say that all are members of the lithe and graceful family Salmonidae—and that the brook trout is the only one that belongs here.

Char are mostly a northern clan, but like so many boreal animals the brook trout makes an incursion south along the Appalachians, reaching down the mountains into Georgia. It requires two things that the Appalachians provide in abundance—pure water and cool summer temperatures. A brookie's comfort range is fifty to sixty-five degrees Fahrenheit, a level that not coincidentally assures abundant dissolved oxygen. Like the flame it sometimes resembles, a brook trout thrives in a charged environment.

When the eastern seaboard was a wilderness, brook trout were apparently common in most streams and fast-flowing rivers, but their lack of caution made them easy to catch, and as the forests fell the streams were deprived of their shade, allowing summer water temperatures to rise to lethal levels. Almost as bad, agriculture (especially the practice, which unforgivably continues today, of allowing livestock to trample the banks of streams) washed a constant fog of silt into the water. The sediment clogs gills and blankets the streambed with mud, robbing the trout of the clean gravel that they use for spawning and choking their eggs. Brook trout, which need water as cold and clear as a winter sky, could not stand the changes, and today the lowland streams are the domain of chubs and suckers and brown trout, whom tepid, murky water suits.

The brook trout were forced back into the mountains and, even there, the bigger streams are denied them. Sport anglers, having laid down their ten or fifteen dollars for a license and another five for a trout stamp, demand that the creeks burst with trout of acceptable size—and if the fish are dumb enough to fall for pickled salmon eggs, scented dough balls, and drowned worms, so much the better. Many streams are waters that become too warm in summer to support brookies (some get so warm that even the stocked brown trout die in August), but many could support natural, wild brook trout populations if they were not flooded each spring with hordes of hatchery fish, which monopolize the limited food and holding stations, as those crucial spots protected from the current are known. Even when stocking is ended, enough of the hatchery fish survive to reproduce, often muscling the less fecund brook trout out of their homes, or (in the case of brown trout) simply eating their young.

Consequently, brook trout are today restricted to the smallest of mountain streams, where food is scarce and a trophy brookie is a whopping seven inches; a ten-incher is something to brag about. Many fishermen, familiar only with these stunted jewels, believe that brook trout are inherently tiny, some sort of natural runt. I once got into a heated and protracted argument with members of a sportsmen's club that wanted to stock a small mountain creek in my county with hatchery fish, on the theory that "decent" trout would get bigger than the natives.

The logical explanation, that the brook trout were undersized because the stream was small, nutrient-poor and suffering from increasingly acidic water, fell on deaf ears. The trout were stocked for several years in a row, and I later caught one, a twelve-inch brown with a jug head and an emaciated body like a hammer handle; a stream that could support a few sublegal natives obviously could not support twice as many fish, all twice as big, and they were starving. The brook trout vanished and so, eventually, did the stocked fish. The last time I could bring myself to search, I found no trout in the stream at all.

Attitudes are changing, fortunately. Many states are actively trying to preserve wild brook trout instead of affording them benign neglect. My home state of Pennsylvania increased the minimum size for all trout from six inches to seven when biologists discovered that wild brook trout generally do not breed until they reach that size, and ceased stocking waters that contain sizable populations of wild trout. Farther south, in the Smokies, where logging and thoughtless introductions of brown and rainbow trout fifty years ago sent brookie populations reeling, managers are trying to undo the damage by building fish barriers at the lower levels of mountain streams, then laboriously electroshocking the upper reaches of the creeks and removing the alien trout.

Greatly complicating matters, however, is the fact that northern brook trout were also stocked in the southern Appalachians. The northern race has merged with the southern, swamping it in many streams to create a mongrel. Concern over this issue is not merely a matter of historical preci-

Fog envelopes the spruces on Clingman's Dome, North Carolina. As tart as lemon juice, such fogs pose a threat to high-elevation forests, and the acid precipitation continues to harm vulnerable streams and lakes.

sion. Recently fisheries biologists have discovered that the southern brook trout is genetically distinct from its New England cousin, perhaps even a species unique unto itself, and without question an irreplaceable slice of the Appalachians' biological diversity—a realization that may have come too late for the native southern strain.

In the 1980s and 1990s, brook trout were drafted into a new role: miner's canaries. Even though they have a somewhat higher tolerance for acidity than most fish, brook trout populations were carefully monitored as an early-warning system for acid rain damage to creeks—in part because the waters they inhabit, flowing over rocks with little in the way of neutralizing limestone, show signs of damage more readily than valley creeks flowing over well-buffered strata.

Acid deposition—produced by emissions of sulfur dioxide and nitrogen dioxides, carried thousands of miles through the air and falling as rain, fog, snow, sleet and dry particles—is a many-faceted evil. Best known for killing fish, amphibian eggs, insects and other aquatic life, famously rendering some lakes and streams all but dead, it has many other insidious effects. It degrades soil, retards the growth of plants and weakens the health of forests. It leaches nutrients out of the soil, while freeing up toxic metals like aluminum. It is an especially potent problem at high elevations, where fog carries the greatest acid wallop, routinely measuring a pH of 2.6, as pickled as lemon juice. This is one reason that high-elevation waterways have suffered most, and in many respects have failed to recover even as levels of sulfur dioxide and nitrogen dioxides have fallen in the past two decades, thanks to more stringent air pollution regulation.

Acidity is a complex threat, as research by the Cornell Lab of Ornithology unexpectedly discovered. While looking at the connection between forest fragmentation and declining wood thrush populations, Cornell scientists realized that the areas with the largest drop in thrush numbers also received the most acidic rain. Why? It appears the answer is calcium, which acid rain leaches from the soil. Wood thrushes, like all songbirds, need extra calcium in summer when the females are laying their eggs, and so they eat land snails, whose shells are rich in the mineral. As soil calcium disappears, so too do snails—and the thrushes suffer eggshell defects that impair their ability to breed.

Many of the pollutants that cause acid rain, along with others like ozone, have also robbed the Appalachians of one of its most fundamental treasures—a clear view. In Shenandoah National Park, visibility has dropped by 80 percent in the summer and 60 percent overall, to a summer average of just eleven miles. Where visitors once could routinely pick out the Washington Monument seventy miles away, the visual range hasn't cracked fifty miles in decades, even under the best conditions. The average visibility in the Great Smoky Mountains has dropped by two-thirds, to just thirty-three miles, and it's the same up and down the East Coast, as anyone with eyes can tell. According to one study, the "median visual range"—the farthest distance the eye can distinguish landscape objects— in the East has dropped from ninety miles under pristine conditions to just fifteen miles.

The good news is that acid precipitation is declining, as the emissions that produce it have been steadily regulated—first at power plants, and more recently from vehicles. But a century of acid damage has so seriously degraded the natural buffering capacity of some watersheds, that even the reduced level of acid deposition today continues to cause many of the same problems conservationists were warning about decades ago.

Acid rain is the slow way to kill a stream. There are quicker, more dramatic methods.

Living as I do on the fringe of the anthracite belt, I've seen my share of dead waters, the victims of mine acid drainage and the other ills of coal mining. Across the street from my aunt's house in Ashland flowed a creek which, when I was a child, mesmerized me with its utter absence of life. It flowed black, carrying a constant load of culm, the fine particles of shale and coal washed down from the great waste banks beside the old breakers. When I stuck my ten-year-old finger in the water, it vanished as though by magic; I'd never seen water so opaque. I hunted along its banks for pieces of shale that carried the fossils of ancient seed-ferns and for "sulfur diamonds," the yellow crystals that precipitated out of the saturated solution of the water on streamside rocks. From a distance, the creek was a Halloween decoration, black water rimmed with rocks stained bright orange from dissolved iron.

Thousands of waterways up and down the Appalachians have been victimized by coal mining, many by the insidious seepage of acidified water from abandoned mine shafts; such streams (if they don't carry a coal silt load) often have an unnatural clearness to the water, which, along with the telltale orange tint of the bottom sediments, speaks of their death. The West Branch of the Susquehanna is one of the biggest casualties and one of the most painful to see, flowing as it does through some of the wildest and prettiest terrain in Pennsylvania and collecting some of the least sullied streams in the region. Its headwaters pass through the bituminous fields, however, where water percolating down through old mines collects a burden of sulfur and iron. Highly acidic, the water dissolves aluminum, manganese and other pollutants before surfacing, a toxic infusion that kills virtually all life. The West Branch is a big, scenic river, but it has an unnatural aqua-blue cast to its deepest pools, testimony to the tortured chemistry of its waters.

Acid-tainted rivers can be reclaimed, although the job is laborious and expensive. The openings that deliver acid must be stopped up or, more often, equipped with devices that feed lime into the seepage to neutralize the acid. More recently, bogs and swamps have been found to be fairly effective at removing particulate matter and buffering acid, and the coal industry—recognizing a cost-effective alternative when it sees one—has responded by constructing hundreds of man-made wetlands.

A wetland improves mine drainage water in several ways. Simply slowing the flow of water through massed stems of cattails, phragmites or a sphagnum moss mat allows some of the pollutants to drop into the bottom sediment. Other chemicals are actively absorbed or altered—manganese and iron by some of the plants, sulfur by bacteria. By the time the water seeps out the far end, it may be substantially free of pollutants and considerably less acidic.

Even that vile creek that flowed past my aunt's house is benefiting from the ministrations of a marsh, this one accidental rather than planned. When a deep mine along one of its tributaries was abandoned in the 1950s and the pumps shut down, water began filling the shafts. The water didn't stop there, however; it eventually filled the old mine and rose to the surface, flooding the small town of Connerton and forcing the relocation of its residents. The buildings were demolished, replaced by a twenty-acre pool of water with a huge culm bank at one end.

Such water should be highly acidic, but the Connerton marsh is an example of what a wetland can do for water quality. The first time I visited the marsh, at the urging of a state hydrogeologist, I was shocked. Not by the thick stands of phragmites reeds, which can tolerate adverse growing conditions, but by the wildlife—wood ducks and mallards, muskrat lodges, water snakes, green frogs, dozens of red-winged blackbirds, even an American bittern, which is a threatened species in the state. Carp rolled in the shallows, and I saw pumpkinseed sunfish and white suckers. Beavers had moved in and dammed the downstream edge of the marsh, raising the water level and expanding the size of the marsh substantially.

The Connerton marsh was no garden spot, mind you. Shattered refrigerators poked above the surface here and there, and we had to be careful not to slit open our waders on broken bottles that crunched underfoot. At the far end, where the water ran over the lip of the beaver dam into Shenandoah Creek, the air had the tang of sewage carried by the stream, and the lowest branches of the gray birches growing along the banks were festooned with bits of plastic and rotted paper. But the marsh had done its job; the pH of the water flowing into the creek was close to neutral 7, instead of the readings of 3 or 4 normal for mine pool water.

The Little Schuylkill River was another victim of mining. It rises in the ridge system north of the Pennsylvania town of Tamaqua and flows south and west for about twenty-five miles, joining the Schuylkill River just as that waterway rushes through the Kittatinny Ridge and into the Great Valley. Like so many coal region waters, the Little Schuylkill carried the effluvium of both deep mines and surface strip mines—acid runoff, coal silt, the whole life-choking mess. The river died completely; a survey in the mid-1960s showed not a trace of visible life in it. Not everyone realized this, however. An acquaintance was driving across a bridge on the opening day of trout season many years ago and saw a fellow hip-deep in the stained water, casting. He stopped to chat, amazed that anyone would waste his time in the river. The angler, it turned out, was from Philadelphia and had just driven north until he found a likely looking place to fish. My friend left him in blissful ignorance, and the locals had a good laugh at his expense.

That Philadelphia fisherman was twenty years early, it seems. In the 1970s, efforts were made to curb both the culm erosion and the acid runoff getting into the Little Schuylkill; some natural bogs and marshes added their abilities to the effort. The river began a remarkable comeback that continues to this day.

Last summer I spent an evening fly-fishing along the Little Schuylkill, something I now do regularly, on a portion of the river that is specially regulated for the sport. All through the late afternoon, a mixed bag of caddisflies came off the water, doing their mothlike dances as they broke free of their nymphal skins and skittered into the air. The trout were there to meet them, splashing up to snatch the clumsy insects before they could become airborne, sometimes jumping completely free of the water to take one from midair. Canada geese, goslings in tow, patrolled the far bank, and wood ducks whipped past me several times with the unmistakable air of being late for an appointment.

I had come through this area for the first time in 1975, paddling a canoe with a high school friend. At that time, the only trout to be found were at the mouths of tributary streams, where the murky river water was somewhat diluted and cooled. We fished, but the only thing we caught were creek chubs, and few of them. When I picked up rocks from the bottom I disturbed thick, gooey black mud and found only some tubifex worms, which are so tolerant of water pollution that they thrive in sewage treatment plant pools.

The difference that twenty years makes is remarkable, although the Little Schuylkill River is still far from pristine. Litter is common and, despite heavily publicized cleanup campaigns, some locals still see the river as a dumping ground, depositing their trash along its banks. Several days of heavy rain will wash coal silt down from the culm banks upriver, staining the water like soot, and its aquatic life is still skewed heavily toward caddisflies, which are forgiving of water pollution; the more sensitive mayflies remain uncommon. Even the trout are on a kind of intensive care program, regularly augmented by state hatchery stockings because there is little natural reproduction.

But such concerns are nit-picking. Ospreys and great blue herons patrol the waters now, and I've even seen migrating bald eagles swooping after fish. On two occasions I've found the distinctive scat of river otters, although I have yet to see the animals themselves. And where I once canoed a damaged, littered river in solitude, now I must now share it with other boaters and fishermen. It is cleaner but more crowded. In all, it's a trade I am happy to make.

In the annual round of seasons, every naturalist observes benchmarks, the moments to which the year is moored. My summer does not begin until the bee-balm blooms, and autumn cannot come until I've seen the first hawks flying south.

Spring, for me, is less a matter of warm days and daffodils than it is of sound. Each year I go out to meet the season on a mild night in a marsh not far from home. The ice and snow, recently and unlamentably gone, have flattened the cattails and sedge leaves into brown hummocks, and black water shines between them in the beam of my flashlight. The western sky still holds its red, but the east is blue-black with night, and the treetops silhouetted against the horizon quiver as a moist Gulf wind breathes through the hills.

Standing motionless in my waders, eyes closed, feeling the cold of the water through the rubber boots and wool socks, I wait for spring to begin its song. I am never disappointed. With the piercing clarity of a boy tenor, a single, piping voice rises from the darkness. Then another, and another, until the night shakes with their songs, hundreds upon hundreds, and my eardrums vibrate so that I cannot hear anything but the hammering *prreeeeps* coming from all sides.

The singers are the tiny chorus frogs known as spring peepers, each no larger than the last joint of my little finger. The males, having left their hibernation sites under leaf litter just days before, migrated through the damp woods to the marsh, setting up their miniature territories beneath the concealing overhangs of dead cattail leaves. The color of dead leaves themselves, with camouflaging streaks of brown crisscrossing their backs, they are for much of the year the most retiring of animals, but spring hangs the males on the horns of a dilemma. They must quickly attract the notice of females in order to mate, while not drawing the attention of predators (for unlike many amphibians, peepers have no toxic defenses).

They solve the problem by flooding the night with their calls. A single peeper raising its voice in isolation would be an easy mark, but a thousand calling at once create the natural equivalent of white noise, so disorienting that it is almost impossible to locate an individual. The mass calling serves other purposes as well. The breeding aggregation makes it easier for the sexes to come together and gives females a greater choice in mate selection.

To mammalian ears a peeper chorus is a confusing blast of sound, but female peepers can discriminate among the assembled singers. They seek out males that are calling slightly faster than their neighbors—males, it turns out, that are somewhat bigger and (as far as researchers can tell) older than those calling more slowly. A male that has survived longer than his neighbors probably enjoys a survival edge that may be hereditary and so would be the best choice to fertilize a female's eggs. Males, for their part, are not passive in the process. They actively defend their territories, each about the size of a serving platter, against incursions from other males. Although one peeper territory looks pretty much like any other to our eyes, there must be aspects to which we are blind that make some more desirable than others.[2] Not all males hold a territory; some, known as satellite males, float through the marsh trying to intercept females—or wait until a dominant male is busy mating with one female, and usurp his singing perch in hopes of duping others.

All across the marsh I can hear the sounds of argument underlying the basic song of procreation. The normal "advertisement" call, with which males attract females, is a sharp, ascending *prreeeep*, repeated about once a second; the whole chorus of males tends to sing in general unison, clamoring for fifteen or twenty minutes, then falling silent for a few minutes as though a conductor had waved his hand. But mixed among the advertisements are aggressive calls, a quicker, stuttering version like someone dragging a fingernail along the teeth of a comb. This call is given by territorial males when they notice a satellite male encroaching on their boundaries, or another male who has taken a singing perch too close for comfort.

Spring peepers are the loudest of the season's heralds (on a still night from my front porch I can hear the clamor from a pond nearly a mile and a half down the valley), but they are not the only singers, nor even the earliest. Even before the snow is fully melted the wood frogs gather, usually in the seasonal ponds that will vanish with midsummer; their tadpoles must race the July droughts, changing from larvae to adults while the water lasts, if they are to survive. Masked like raccoons and quacking

His balloon-like vocal sac inflated, a male spring peeper fills the April night with his ear-piercing courtship calls.

like ducks, the wood frogs are unusual also in that they conduct their rituals of courtship and mating in broad daylight, perhaps to take advantage of the weak March sun's warmth. There is none of the careful, territorial spacing of the spring peepers among wood frogs; the males simply float on the surface, legs akimbo and eyes bulging, inflating their vocal pouches in front of their forelegs. Any female—any *thing*—that swims or floats by is fair game, grabbed from the rear in the grasp known as *amplexus*, in which the male hugs the female behind her front legs, hanging on until she extrudes her mass of eggs, which he then fertilizes.

Wood frogs emerge first to sing, followed a few days or a week later by the peepers. The rest of the choir waits for the more moderate days of real spring, but by mid-April the marshes in my area are pulsing with a variety of calls—the weird, musical buzz of American toads, the plunk of green frogs, the resonant snores of pickerel and leopard frogs. Later still, in May, come the bass rumblings of bullfrogs and the squirrel-like trills of gray treefrogs.

Salamanders breed in many of the same springtime wetlands as frogs and toads, but they lack any talent for musical expression, and so are easily overlooked. Many species have a tolerance for cold that exceeds even a wood frog's and complete their breeding before most human naturalists think to look for them. This is particularly true of the family known as mole salamanders, which, despite their name, are among the loveliest of Appalachian creatures. The spotted salamander is the most widespread and common, found everywhere in the mountains from the Gaspé south, although it is absent from the highest conifer zones. Averaging six to eight inches long, blue-black with a dozen or more large, lemon-yellow or orange spots down its back, this is a spectacular animal that hides for most of the year, burrowing through the soil and sheltering beneath dead logs and pieces of bark.

On the first rainy night of spring—usually a raw, miserable soak when the air temperature is barely ten degrees above freezing—male spotted salamanders emerge from the ground and commence a slow-motion migration through the forest to their breeding pools. On the next rainy night, the females make the same trek. Some ponds attract hundreds, even thousands, of salamanders, and they stream to water like commuters into a train station, instinctively following gravity downhill. For someone who has never heard of amphibian breeding concentrations and wouldn't know a spotted salamander from a banana, I imagine it is a shock to come around a bend on wet night and find the road alive with squirming, eight-inch, polka-dotted animals, moving as though they're on a mission. Which, of course, they are.

In fact, roadkills are a problem for mole salamanders in some areas, where development has encircled their breeding pools. Even more than frogs and toads, salamanders are tradition-bound creatures, drawn back season after season to the place of their birth, then departing again until the following year. One can only guess how it must feel for the salamander to emerge from the earth to discover that the wet, loamy woodlot through which it used to migrate is now the Red Oak Farms housing development, and the seasonal pool that used to stand among the oaks from first thaw until midsummer has now been graded for, say, the community tennis courts. Just because a wetland isn't permanent doesn't mean it isn't important. (In fairness, it must be said that some communities have tried to protect their salamanders by briefly closing roads, posting "Salamander Crossing" signs and even—taking a lead from Europe—building guide rails and underpasses to direct the migrants through culverts.)

Those salamanders that navigate the twenty-first-century roadblocks and come together for their ice-water orgy complete their breeding cycle in just a day or two. The males and females circle each other in slow courtship dances, the male periodically nuzzling the female with his chin. At length he deposits on the bottom a spermatophore, a tiny gelatinous cone capped with sperm; the female picks it up with her cloaca, and fertilization occurs within her body before the eggs are laid—a less risky process than the external fertilization of frogs and toads. When the eggs are finally laid, anytime up to forty-eight hours after mating, the female masses them on submerged branches so that they form a glob the size of a large orange. Soon after, again on a damp night of rain, the adults leave the pond to resume their subterranean lives.

The hazards of modern life that face amphibians go beyond the wheels of pickup trucks. Beginning in the 1980s, scientists began to notice an alarming and mysterious disappearing act, as once-common species declined or vanished altogether. The hardest-hit species have been the anurans (as frogs and toads collectively are known) but there have been decreases in salamander populations as well.

Researchers scrambled to find a cause, and there were many potential contenders, ranging from environmental toxins like pesticides and herbicides, to increases in ultraviolet radiation, to climate change and habitat loss. No single explanation fit all the circumstances, though. Because many amphibians breed in standing water, acid rain was an early suspect, especially for species like mole salamanders and wood frogs that breed in pools of rainwater or snow melt. Yet many of the population collapses were in regions with little or no acid deposition. Global warming was a possibility, because temperature changes can be accompanied by changes in rainfall and humidity—two crucial factors to moist-skinned amphibians.

None of those possible causes, however, explained the terrifying speed with which the population collapses occurred, even in seemingly pristine environments. The golden toad of Costa Rica suffered a 99 percent population fall in just two years, and soon thereafter was entirely extinct. And that dazzling, beautiful species was only among the first; in the past two decades, more than a hundred species of amphibians have disappeared around the world, while hundreds of others teeter on the edge.

The cause? Scientists finally isolated a fungus, *Batrachochytrium dendrobatidis* (also known as *Bd* or chytrid) that fatally thickens the skin of an infected amphibian. No one knows if it was always present in amphibians but has been triggered by some environmental factor, or recently jumped from another taxa; it is common in African clawed frogs, popular in the pet trade and research alike. Wherever it started, and however it spread, chytrid is burning its way through amphibian populations in essentially every corner of the world in what's been described as a wave of death. In terms of geographic sweep and number of species effected, chytrid is easily the worst wildlife epidemic in history, and it is on a pace to wipe out half the world's amphibians within a human lifetime.

Nor does chytrid operate in a vacuum. All those other potential suspects that scientists initially examined, from chemicals to climate, are having their own cumulative, negative effects on amphibians—which with their moist, absorbent skins are exquisitely sensitive to contamination and environmental change. If there is any good news it is that, so far at least, the impact of chytrid on Appalachian amphibians has been relatively light. Recent surveys, for example, found little evidence of the fungus among salamanders in headwater streams of the southern Appalachians—a relief, given the globally significant diversity of salamanders there. But the looming threat of a salamander-specific chytrid now rampaging across Europe – and as close as one infected specimen coming in through the exotic pet trade – has herpetologists terrified.

All these are gloomy thoughts for an April night, when the peepers are shrilling their songs of sex to the marsh. For the moment, I try to push them out of my mind and concentrate on the pulse of the sound, the way it makes the bones in my head resonate and my ears buzz. Spring is back, and her chorus is in full, passionate voice.

Hardly anyone pays attention to mussels. They haven't the bright colors of a brook trout or the ringing song of a peeper, and they live their lives out of sight, half-buried in the gravel of river bottoms.

Even many naturalists who appreciate the diversity of the Appalachians' wildflowers or salamanders don't realize that the region hosts one of the world's great mussel faunas. Or at least it did; these unassuming bivalves are vanishing at an unprecedented rate, the victims of dams, water pollution and other evils. They are dying out, not just in ones and twos, but in whole assemblages of species.

Their loss should concern us for all the good and usual reasons—the inherent value in any unique organism and the way it fits into the functioning whole, and what its potential loss says about our treatment of the world in general, and this small slice of it in particular. But I will mourn the mussels for a less serious, but I think no less valid, reason. For they are the most whimsically named group of creatures in North America.

Say some of the names out loud: Shiny pigtoe. Pimpleback. Monkeyface pearlymussel. Rough rabbitsfoot. Dromedary pearlymussel. Snuffbox. Purple wartyback. Pink heelsplitter. Just try to say "pink mucket pearlymussel" without smiling. You can't do it.

I do not mean to trivialize the mussels' plight by this focus on what we humans call them; rather, I find it one more reason to rail against their destruction. For all their excessively retiring habits, they brought out a streak of fancy in us, and we should celebrate them for that, along with the more mundane reasons of ecology.

The East has the world's richest mussel fauna, at three hundred species nearly a third of the global total, and much of that diversity inhabits the streams and rivers of the Appalachian Mountains and Plateau. Nearly four dozen species have been found in just a few hundred yards of a single river, triple what occurs in the whole of Europe.

Just as different species of warblers have specialized to hunt or nest in different levels of a single tree, so have mussels fitted themselves into their own narrow niches. Each species prefers a certain mix of water depth, current flow, substrate and temperature, divvying up the river among themselves. But their specialization goes further. Mussels go through a brief parasitic stage as larvae hitchhiking on fish, and each variety has only a single host species (so far as we know—the host fish have been identified for only one-sixth of all North American mussels). To dupe the fish into coming close enough to pick up the larvae, the adult female mussel may grow an artful extrusion of its fleshy mantle, which it waves in the current. The effect is startlingly like a minnow. The fish, thinking it's getting an easy meal, attacks—and the female releases a cloud of larvae in its face. Over the course of her fifty-year lifespan, she may play the trick thousands of times.

The waters of the Appalachians' west slope are also home to a striking diversity of fish—especially one family, the darters, which after minnows are the most varied group of fish in North America. In the 1970s one of them, the snail darter, blazed to notoriety when its status as a federally protected threatened species held up the infamous Tellico Dam on the Little Tennessee River, thought to be the only home for the darter (other populations were later found elsewhere). The arguments in favor of the dam were, at best, shaky, but the darter's interests were eventually ruled secondary to economic considerations, and the Tellico Dam was built. It drowned the Southeast's best trout stream for no particularly good purpose, and "snail darter" has since become synonymous with insignificance.

Darters have always struck me as having more personality than the average fish. They are small, usually between three and five inches long, but their shape is distinctive and instantly recognizable—a long body with a blunt head, two high dorsal fins over the back, and enlarged pectoral ("chest") fins, on which the darter supports itself while hopping across the bottom. Its eyes bulge from their position high on the fish's head; the whole effect is froglike and gently comical. Over the years I've kept many of the common tessellated and Johnny darters in home aquariums, and unlike the rest of the fish in the tank, which tend to drift vacantly about, the darters always show an interest in what is going on out on my side of the glass. I'll glance up from my work to see one or two lined up at the front of the aquarium, peering intently at me.

Darters are also, as a group, among the most colorful of North American fishes. A male of the appropriately named rainbow darter is typically gaudy, with a red body striped with blue bars, colorful dorsal fins (red and blue in the front, orangish red in the rear) and a large blue anal fin with red streaks. Other males of the same species sport bodies that are cobalt blue, with each scale carrying a spot of crimson. Rainbow darters look like escapees from a tropical fish store, but they are common in streams and small rivers from the Appalachian Plateau to the Great Lakes region.

While some species, like the Johnnys and tessellateds in my home tank, are simply drab brown, most are splashed with red, orange or blue, and many are standouts. The male sharphead darter of the Iron Mountains region, where Virginia, North Carolina and Tennessee meet, is a blue-green wonder with turquoise fins; males of the tiny Tippecanoe darter are a deep orange, with blue chests and indigo bands around the base of their tails.

Like the mussels, whose center of diversity they overlap, the darters have become localized specialists. Some require fast water with a cobblestone bottom; others, the same current speed but a sandy substrate or one broken by large, emergent boulders. Others need quiet pools with thick aquatic vegetation or swift, deep holes with a gravel base. Almost all are creatures of clear, fast and relatively shallow water, which is why damming, dredging, siltation and other river alterations are so damaging to their survival.

Many species exist only in particular river systems—even only in individual rivers. The Conasauga logperch, a largish darter on the federal endangered species list, is found only in rock-strewn, fast-flowing sections of the Conasauga River on the Georgia–Tennessee line. The Conasauga River is home to another endangered darter, the amber darter, which is restricted to a few sandy riffles in that waterway (another small population inhabits the Etowah River basin farther south in Georgia). Frequently, even those darters with fairly wide geographical ranges may be scattered widely within their boundaries. The longhead darter, which is found along the Appalachians' west slope from southwestern New York to the Smokies, is so localized and rare within that huge area that it, too, is being considered for the endangered list.

Specialization is always a trade-off for any organism. As long as conditions remain stable, specialists may enjoy a competitive advantage over ecological generalists, which can tolerate many conditions but are rarely equipped to exploit any one situation to the fullest. But when the playing field is altered—by a natural change like glaciation or a man-made cause such as a dam—the specialist may find itself painted into a corner.

Because mussels are so specialized, they are among the first to suffer when a river system is altered by humans. Large-scale damming, like the Tennessee Valley Authority's glut of reservoirs, has killed many; siltation, acid mine runoff, agricultural chemicals and industrial pollution have hammered those in the remaining free-flowing rivers. Over-collection for the mother-of-pearl button and freshwater pearl trades devastated populations in the past; today, poaching for the shells, which are used in Asia as the seeds for cultured pearls, is a rising threat.

Overall, the toll has been worse for mussels than for any other class of animals in North America: twelve species extinct, forty-two listed as threatened or endangered, another eighty-eight so badly off that they are under consideration for federal protection. That's nearly half of all the mussel species on the continent.

The newest danger, oddly enough, is another mollusk: *Dreissena polymorpha*, the Eurasian zebra mussel. It first showed up in Lake St. Clair, between lakes Erie and Huron, in 1988, probably carried there as larvae in the ballast water of a tanker several years earlier (unlike most freshwater mussels, the zebra's larvae are free-swimming). "Prolific" does not even begin to describe the zebra mussel, which started a conquering sweep that would have done the Mongol Horde proud. Within just a few years, it had so monopolized Lake St. Clair and parts of Lake Erie that native mussels were gone; the zebra mussels had consumed so much plankton and algae that the lake water cleared noticeably. The zebra mussel continues to spread at a frightening rate, and now occupies the entire Great Lakes basin east to the Hudson and Chesapeake, south along the Mississippi to Louisiana, and in pockets as far west as California. They have gotten a lot of attention, if only because they step on human toes, too. At peak densities, zebra mussels may be packed at a rate of one hundred thousand per square yard, and can choke off power plant and water filtration intakes. A larger, related species, the quagga mussel, is also spreading out from the Great Lakes, compounding the problem.

Native mollusks simply can't compete; the tiny zebras just use them as another anchoring point and smother them beneath layers of striped shells, meanwhile scarfing up so much of the available food that they threaten whole aquatic food chains. Biologists have a sick dread that, as the zebra mussel begins marching through the rivers of the Appalachian Plateau, it will sweep away the products of millions of years of evolution and specialization. This is not certain; zebra mussels like still or slowly moving water, and many native species prefer faster currents, but hardly anyone is hopeful. The recent discovery of a second species of zebra mussel, this one tolerant of an even wider range of water conditions, may make the point moot. For the pigtoes and riffleshells and heelsplitters the future—which was already bleak—is looking steadily worse.

The usual image of Appalachian waters is of a mossy stream tumbling over itself, full of brook trout and mayflies, but, for much of its flow, the Hudson River is also Appalachian. Although born in the Adirondacks, the river cuts a wide valley down the eastern bootheel of New York, skirting the Catskills and the Shawangunks to the west and the Taconics to the east. Below Newburgh it cuts through the steep, sudden jumble known as the Hudson Highlands—actually the Reading Prong, a narrow waist of granite hills that joins the Blue Ridge system to the south with the Berkshires and the Green Mountains to the north.

Flush with runoff, Curtis Creek spills through a deep valley in Pisgah National Forest in the mountains of North Carolina.

It is a dramatic landscape, with Bear Mountain, Storm King Mountain and other peaks hanging over the Hudson, walling it in with cliffs and sheer, wooded slopes. Yet what makes this conjunction of river and mountains most unique isn't immediately apparent. You have to take your eyes off the scenery, however impressive, and look at the details.

The trail that winds from the small visitors center at Constitution Is-land Marsh Sanctuary, on the east shore of the Hudson a few miles south of Newburgh, looks like any other trail in the central Appalachians. It rises and falls over outcroppings of conglomerate and is shaded by a forest of chestnut oaks, where lady's-slipper orchids grow in spring. At length the trail comes out to the marsh, where a boardwalk cuts through acres of cattails, jewelweed and scattered cardinal flower.

This is where the Hudson's unique place among Appalachian waterways can be appreciated. Even though Constitution Island is nearly sixty miles from the ocean, this part of the river is actively tidal—a brackish environment where ocean and mountains meet. Within a few yards you can go from a world of white-tailed deer and copperheads to one of crabs and barnacles. It is the only place south of Canada where the Appalachians and the Atlantic intersect.

At Constitution Island, the tide rises and falls an average of four feet, but the tidal influence extends much farther upriver—all the way to Troy, some 154 miles from the Hudson's mouth. In addition to the pugnacious blue crabs, the estuary is home to several species of ocean migrants that enter it to spawn—herring, striped bass and others—along with freshwater fish like largemouth bass, sunfish and white catfish. Two species of killifish mingle in the Hudson estuary: banded killifish, a common freshwater species tolerant of some salt, and mummichog, a coastal species that can tolerate freshwater.

At low tide (indeed, at any time) the Hudson estuary is teeming with life. Great egrets, great blue herons and the occasional snowy egret hunt the mudflats left by the receding tide, and I've seen absolutely spectacular flocks of tree swallows swirling in to roost in the evening by the thousands. The estuary is home to many snapping turtles, including one caught a few years ago that could be a new world record—a monster with an upper shell nearly twenty inches long.

A first-time visitor, seeing the riot of life and the wild-looking mountains standing over the river, can be forgiven for thinking that this is a little slice of protected paradise. In truth, the Hudson was a sorely abused river, so polluted by the middle of this century that it was frequently referred to as an open sewer. Frankly, that comparison may have been unfair to sewers, because the Hudson also carried a nasty brew of heavy metals, PCBs and other invisible toxins.

Pollution control laws have worked wonders, but while the Hudson River estuary is cleaner now than it has been in decades, it is still a long way from pristine. Blue crabs are an important commercial crop (the Hudson bids fair to unseat the Chesapeake for title of America's richest blue crab fishery), but the state advises eating only one a week, and none at all for women under fifty or for kids, because the crustaceans carry such a heavy load of PCBs and heavy metals.

At one time, virtually all the great rivers flowing from the Appalachians to the Atlantic carried an upstream flood of migrating fish—American shad, blueback herring, alewife, American eels, striped bass and, in New England and Canada, Atlantic salmon. With the exception of the eels, these are anadromous fish, spawning in freshwater but spending most of their lives at sea (eels, which are termed catadromous, spawn in the ocean but mature in freshwater).

The Susquehanna River in Pennsylvania once supported the East's greatest shad run, a springtime event when the wide, shallow river choked with millions of American shad—"bucks," or males, weighing three to five pounds, and "roes," as the gravid females are known, weighing as much as eight or nine pounds. Generally the bucks arrived first, charging up from the Chesapeake not long after ice-out, with the roes following upstream in April and early May. They spawned in the mountain tributaries and the chilly pools of the headwaters as far north as New York, then turned tail— emaciated and worn by the months-long effort—and swam with the current back to the sea to rest and feed through another cycle.

On the Susquehanna, as on so many rivers, dams killed the anadromous fish runs—first, in the early 1800s, dams for the newly built canal system, and in the early 1900s, massive hydroelectric dams like Holtwood and Conowingo. The Susquehanna now has no fewer than six dams between its mouth and the juncture of its east and west branches at Sunbury, most of them clustered at the southern end nearest the bay—thus eliminating even a relict run of fish. The shad vanished, and the remains of old stone fish traps and weirs, which you can still see from the air on the bottoms of some tributaries like the Juniata, are the only reminders of their former magnitude.

Actually, the Susquehanna again has a shad run of sorts. Over the years, the state and federal governments have badgered the electric utilities that own the dams to install fish traps—sophisticated variations on the old stone pens and guide walls of the nineteenth century. Shad were stocked in the river, and when their offspring tried to follow their instincts upriver, they were corralled, netted, and trucked up above the dams. Some were released to spawn naturally, but most were held in hatcheries, their fry later released to migrate to the sea.

The obstacles are tremendous. Not only must adults going upriver battle dams, but the young shad, which migrate downstream the following autumn, must contend with them on the seaward journey. The fry are presented with two equally deadly alternatives: being sucked through the power-generating turbines or being swept over the spillway and churned to a pulp below. An estimated half are killed on the downstream migration, and there is no feasible way yet to get them around the hazards.

Hauling shad across the countryside in trucks is no one's idea of a terrific solution, and conservationists have been pressing for a more permanent alternative. Sadly, as long as the dams are in place, the shad will need considerable human help. After decades of negotiation, several utilities that own dams on the lower Susquehanna installed fish lifts, which capture adult shad and herring and, like a watery elevator, transport them up and over the dams. This has restored a limited shad run to the lower Susquehanna. In 2013, roughly thirteen thousand shad were lifted over the first dam, Conowingo—although barely two hundred passed the fish ladder at York Haven Dam, the fourth barrier some forty miles upriver. It's progress, but we're still a long way from the days when shad swarmed by the millions up the river.

There are rivers, the Hudson and the Delaware among them, that don't have huge dams near their mouths, and so have retained some of their anadromous fish runs. Here pollution is often as great an obstacle as any physical structure; in dry years when river levels are low on the Delaware, the "pollution block" of tainted, oxygen-poor water in the Philadelphia/Camden/Trenton corridor keeps many shad and herring from pushing upstream.

Most years, however, the shad surge ahead, and the word gets out through the informal grapevine of anglers: Lots of bucks at Martin's Creek. A few roes coming through Sandt's Eddy and Upper Black Eddy. Anything at Shawnee? No, not yet. But soon.

Above Easton the river and the ridges meet, and the shad come back to the Delaware Water Gap with the wildflowers of late April and early May. The fishermen are waiting for them. Shad do not feed in freshwater—and they are planktonic predators in the ocean—but something makes them strike at brightly colored or shiny objects, and anglers perch along the banks or sit patiently in boats, drifting small, painted lead lures known as darts.

Shad are powerful fish, with a deeply compressed body in profile and a slim, current-splitting shape when viewed from the nose. They are covered in silvery scales with a hint of purplish iridescence, and the only markings are a few dark spots along the shoulder. The mouth is large and folds out like a funnel, presumably the better to catch plankton. They lack the slim grace of a trout or salmon, but don't be fooled; when a shad hits a fisherman's dart, the result is explosive.

The first shad I hooked, a five-pound buck, took nearly fifteen minutes to land. Twice I had it up to the boat, only to see it roar off again with seem-

Trillium brighten the mossy banks of a creek high in the mountains—the kind of small, always-cold waterway that today provides refuge for native brook trout.

ingly undiminished energy, and I lost count of how many times it leaped from the water—an eerie sight, since there was a thick, low mist drifting down the Delaware that morning, and the fish simply appeared above it, as though the fog was spontaneously generating silver missiles. I have caught fish much larger, but I have never hooked one that fought as well as a shad.

Because of their power on the line and their tendency to jump, shad are often referred to as "poor man's salmon." The phrase refers to the Atlantic salmon, *Salmo salar*, which is routinely called the king of fish. For once, routine may have it right.

Salar—the name means "leaper"—is grace incarnate. Unlike Pacific salmon such as the sockeye, it does not undergo grotesque physical changes when it enters freshwater to spawn; the Atlantic deepens in color and the males grow a hooked underjaw known as a *kype*, but it retains its perfectly streamlined shape. Atlantic salmon are big (many average fifteen or twenty pounds, with record-breakers of more than eighty pounds), and they fight with strength and reckless, airborne abandon. There is no greater freshwater quarry for a fly-fisherman.

Atlantic salmon are found across the northern rim of their namesake ocean, from Portugal to New England and east to the Barents Sea in Russia. In the United States the species may once have been found as far south as the Delaware River, although more conservative experts place its southern limit at the Connecticut and Houstatonic rivers in Connecticut. In any event, the industrialization of New England, with its pollution and dams, wiped out the salmon runs except for a few small reminders in Maine. Only in Atlantic Canada did the Leaper continue in good numbers, and even there it was beset by increasing commercial fishing and river degradation— first by logging, and more recently by diseases and parasites associated with commercial salmon farms, which sit like pathogen factories at the mouths of many historic salmon rivers, infecting the dwindling wild runs.

In the 1970s fisheries managers tried to undo some of the old wrongs. Fish passageways like lifts and ladders were designed and built, and salmon eggs were raised and the fry stocked. Most of the effort centered on the Connecticut River, which boasted the heaviest Atlantic salmon run in the United States until it was dammed in the 1760s. At first the results were dismal—the fingerlings showed no desire to migrate, a result of using salmon eggs from distant rivers instead of the closest Maine stocks and raising them in hatcheries like domestic trout. But over the next twenty years the managers got better, choosing their egg supplies with genetics in mind, providing a more nutritious diet and housing the fish at lower densities so they were in better physical condition when released into the river.

The payoff came on Halloween, 1991, when a Connecticut fisheries biologist discovered two Atlantics spawning in the Salmon River, a tributary of the Connecticut about twenty miles from the ocean. The fish each weighed about eight or ten pounds and had built three redds, the saucer-shaped gravel depressions in which salmon lay their eggs.

It was the first time in more than two hundred years that salmon had spawned in the Connecticut basin, and the fish had bucked long odds to earn that honor. Clip marks on their fins showed they were two of the nearly forty-five thousand young salmon stocked two years before. Of that number, only nine were known to have migrated up the Salmon in spring, but low water all summer kept them from using the fish ladder for the final ascension around a dam to the spawning area.

Unfortunately, the Connecticut River salmon restoration never really took hold. Despite massive efforts to strip, fertilize and hatch the eggs of returning salmon, thus pumping a million or so salmon fry into the river every year, the returns of adult salmon remained painfully low. The young salmon survived in feeder streams well enough, but most simply never returned from their years at sea; perhaps victims of overfishing in the North Atlantic, perhaps of climate change. After just eighty-nine adult salmon returned in 2013, the federal government pulled the plug on artificial support for the Connecticut fish.

Yet the picture has been brighter for anadromous fish in general; even on the Connecticut, American shad runs of up to a million fish have been recorded in recent years. And Americans have finally gotten serious about restoring free-flowing rivers, especially in the Northeast. In 1999, the Edwards Dam on the Kennebec River in Maine was breached, opening up tributaries like the Sebasticook River, where some 2.7 million alewives now spawn, but where none existed two decades earlier.

Farther up the coast, two major dams blocking the lower Penobscot River have been removed, and fishways are planned for remaining barriers, opening up much of the river to eleven species of ocean-running fish—including the Leaper. Once, the Penobscot supported tens of thousands; the first English to row up its waters described a "great store" of salmon there.

I have always wanted to catch an Atlantic salmon on a fly, but because I am not wealthy man—and because salmon fishing is among the priciest of sports—this always seemed unlikely. Still, one can dream, and whenever I have a chance, I like to stop beside a salmon river and let my mind wander.

I do a lot of this sort of daydreaming in Newfoundland, where the Long Range Mountains shelter many salmon rivers—waters that, if not as famous as the Restigouche or Miramichi in New Brunswick, nonetheless lure the Leaper back from the ocean each year, especially now that commercial salmon fishing on the high seas has been curtailed. I buy a fishing license, but the standard trout permit is good only on those rivers not known for their salmon runs; for five dollars you are free to fish for brookies or sea trout, but if you want salmon you must plunk down a much greater sum, along with the services of a guide. So I stick to the trout streams, knowing that there are at least a few salmon around to keep my pipe dream alive.

Sometimes even the trout snub me. I had been fishing all morning one day last year, hopping from creek to creek like a dilettante, switching flies with the same manic intensity, from a hare's ear fished wet to a dry Adams to a fox nymph to a marabou streamer. There were trout enough—runts that fought for the privilege of biting my barbless hook, but the few larger fish I could see against the undercut banks were lethargic and uninterested.

I finally found myself on a somewhat bigger stream close to camp, and since evening was coming on, I decided to fish from the highway bridge down to the ocean, a distance of a hundred yards, then cast into the surf until dark, hoping to catch a sea-run brown trout coming back to spawn.

Here again, the stunted brookies were as annoying as the blackflies that crawled around my ears, smacking the fly the instant it landed on the water and so monopolizing it that I had little hope of anything worthwhile getting a chance to see it.

Brook trout are not known as fighters, so I was surprised when yet another small brookie smacked the lure—I had switched to a high-floating elk hair caddis—and took it airborne. The fish leaped clear of the water three times, something I've never seen a brook trout of any size manage. I quickly pulled it in (there was no fight, since this was barely a six-inch fish), wetted my hand, and gently eased it from the water. It wasn't a brookie; the fish had a silvery glint, with the dark blotches along each side known as parr marks. Nor did I think it was a young brown trout, for the tail was rather deeply forked, and there were relatively few red spots between the parr marks.

I could not be sure then, nor am I certain even now, having studied the field guide illustrations, but I believe the fish I held in my hands was a young Atlantic salmon, almost ready to leave fresh water and head for the sea. I slipped the hook carefully from the corner of its mouth and returned it to the water—then did a foolish little jig on the shore, clumping in my chest waders, in honor of my first salmon on a fly. There was no one to see me but the gray jays, and they, thankfully, kept their opinions to themselves.

Common Raven

Chapter 5

KEEPING FAITH
WITH THE NORTH

The rain had fallen all night, a pounding onslaught that shook the tent with its vehemence, and turned my campsite in Gros Morne National Park into a shallow pool more suited for salmon than for people. But by morning the overcast was clearing, blown by a ferocious west wind coming off the Gulf of St. Lawrence that smeared the clouds to mist against the Long Range Mountains.

I drove south, climbing into the hills from the East Arm fjord, a long tongue of the ocean that reaches deep into the Newfoundland mountains, where black-backed gulls loafed on sandbars exposed by low tide. The evidence of the night's rain was everywhere, in streams choked to overflowing and in miniature cataracts that leaped suicidally from road cuts and cliffs.

When I parked at the trailhead for Southeast Brook, I could already hear the roar of the falls a quarter mile away. The day before I'd been disappointed by the low water, barely wetting the long, rocky drop. But now the stream had gathered water from the farthest reaches of the deep, glacial valleys at its head, and when I reached the small overlook beside the edge of the falls, peat-stained water was pouring violently over the lip, smashing itself to white foam far below.

While I sat staring at the falls, a movement caught my attention above and to one side—a black shape hanging in the wind. A raven floated down the valley toward me, carefully scrutinizing me as it came. Just across the falls, it landed on the tip of a dead spruce and shook its feathers, like a man settling a jacket on his shoulders.

I've always been drawn to ravens. Some birds, however pretty, have a tendency to fade into the scenery (the great bird artist Francis Lee Jaques once remarked, "The difference between warblers and no warblers is very slight."). But ravens are different; they have a presence about them—a sense of personality, a panache. The difference between ravens and no ravens is tremendous.

This one sat in profile, the wind rustling the shaggy feathers of its throat, giving it the patriarchal, bearded appearance so typical of the species. It watched me with an eye at once casual and calculating, for ravens are believed to be among the most intelligent birds in the world. I'm sure that something very close to real thought—perhaps even reason—was going on behind those shiny black eyes, but I have no way of even guessing what a raven's thoughts might be.

We watched each other for several minutes, as the boom of the waterfall began to make my ears buzz. Then the raven shrugged its wings open, leaned forward and took off, dropping into the void beside the falls. It wheeled—a big, black bird that looked more like a raptor than a crow, its closest relative. The wings were wide and plank-like, the tail long and wedge-shaped, the head prominent, ending with the massive bill. It handled the wind with as much authority as any eagle.

Ravens are emblematic of the North, but like so many northern species, both animals and plants, they have a range that extends far south in the Appalachians, forming a gently curving prong on the field guide range map that dips down into Georgia. The presence of these northerners in a more southerly world is one of the charms of any mountain range, particularly the Appalachians.

Humans, the great pigeonholers, have long tried to impose systems of regimentation on the wild world. In the 1890s, a biologist named C. Hart Merriam described the neat divisions of natural communities on western mountains—the way in which animal and plant associations change at predictable rates with increasing altitude, in the same way they change the farther north one travels. This concept of life zones, as they became known, was eventually applied by Merriam to the whole of North America. Under his system, there were three major regions containing seven life zones. From north to south, they ran like this: The Boreal Region contained the Arctic, Hudsonian and Canadian zones; the Austral Region contained the Transition (or Alleghenian), Carolinian and Austroriparian zones; and the Tropical Region contained the Tropical Life Zone. The Appalachians ran almost the whole gamut, from the Arctic and Hudsonian zones in extreme northern Newfoundland to the Carolinian in the south.

Merriam's system was the first comprehensive classification of the continent's natural communities, but it was far from perfect. For one thing, Merriam used temperature gradients as a foundation for his system, noting that plant communities in western mountains generally change with each four-degree (centigrade) alteration in average temperature. His continental life zones, then, were actually climatic zones.

That's fine as far as it goes, but anyone who has traveled in the Appalachians knows that Merriam's life zones were the most sweeping of generalities. The Transition Zone across the central Appalachians, for example, is really a diverse mix of forest communities—oak woodlands on the low, dry ridges; northern hardwood forests at higher altitudes or on north-facing slopes; montane conifers like red spruce above four thousand feet; white pine on the flats; hemlock in damp, cool ravines; and so on. It is the farthest thing from homogeneity.

Merriam's system is rarely used today. On a continental scale, ecologists talk of biomes (another label for natural communities), with the Appalachians in the eastern temperate deciduous and coniferous forest biomes. Within these two vast arenas a number of major forest types are recognized—the boreal (northern conifer), northern hardwood, oak-hickory and mixed Appalachian forests.

And within these divisions are still more precise ecological units: boreal bogs, northern swamp forests, Appalachian cove forests and so forth, each with its own characteristic mix of animals and plants—the indicator species.

A raven is not an indicator species. Although it is most closely associated with boreal forests, it is found in other habitats, too, from bogs in central Maine to remote oak ridges in Pennsylvania and high mountains as far south as Georgia. But to me, the raven is the heart and soul of the north and the standard-bearer for Merriam's old Canadian Life Zone, which snakes down the spine of the Appalachians.

Ravens are corvids, members of the bird family that includes crows, jays and magpies—all intelligent birds with a reputation for creative puzzle-solving. "The raven is one of our most sagacious birds, crafty, resourceful, adaptable, and quick to learn and profit by experience," Arthur Cleveland Bent said of the species. My favorite corvid story, however, involves not ravens but Swedish hooded crows, which learned to recognize the red flags on ice fishing rigs as signals of a hooked fish. The birds grabbed the fishing lines in their beaks, walked backward a short distance, then walked forward on the lines to prevent the fish from dropping back into the water. A few repetitions, and the fish was landed.

There are many stories of raven intelligence, but anecdotes are not proof, and until recently, there was no experimental evidence that these big birds were capable of intelligent thought. Zoologist Bernd Heinrich, who has studied Maine's ravens, devised a series of tests to settle the issue. Heinrich theorized that if ravens were truly intelligent, they should be able to complete a task with many simple, separate steps, in which none of the separate steps alone brought any reward. In other words, the ravens would have to size up the situation and plan a complete course of action, rather than learning by trial and error.

He didn't say whether he had those Swedish crows in mind, but the test was similar. In an aviary housing wild-caught and captive-reared ravens and American crows, Heinrich suspended pieces of meat from long lengths of string tied to a perch. The crows, while interested in the meat, couldn't figure out that they needed to pull the string up repeatedly, then stand on the slack. Four of Heinrich's five ravens, on the other hand, performed the task successfully on the first try. Even more startling, the ravens never tried to fly off with the meat, as the crows did when presented with a tied hunk. The ravens' behavior suggests an understanding that the string would impede their flight.

Ravens need their brains because they are opportunists rather than ecological specialists and must be able to exploit whatever fortune and the environment bring their way. They are scavengers, for the most part, but it is fair to say that there is almost nothing a raven won't eat if presented with the chance. Although carrion is the mainstay of their diet, ravens have been known to catch a variety of small mammals, reptiles, amphibians, young birds and eggs (particularly around seabird colonies), fish, mollusks, worms, insects, crustaceans, fruit, berries, nuts, field crops, picnic scraps— in short, anything even remotely edible. Nineteenth-century naturalist Thure Kumlien even reported seeing a pair of ravens hunting a seal pup cooperatively—one lying down to the breathing hole, cutting off its escape, the other killing the seal with blows to the head.

Europeans brought a distaste for ravens with them to the New World, for in Europe, the black birds were associated with battlefields and the gallows, and they were hated by shepherds who believed the birds had a taste for sheep. Horace Kephart, who lived in the Smokies for the first thirty years of the last century, reflected that view when he wrote about ravens in *Our Southern Highlanders*:

"As is well known, ravens can be taught human speech, like parrots; and I am told they show the same preference for bad words—which, I think, is quite in character with their reputation as thieves and butchers. However,

I may be prejudiced, seeing that the raven's favorite dainties for his menu are the eyes of living fawns and lambs."

The raven's nesting season begins quite early—in late February in the central Appalachians, when the snow and sleet still fly in the mountains. Ravens usually choose cliff sites that offer shelter from above and from the prevailing winds, building a thick, messy mound of sticks, lined with moss, grass or hair, although the nest may be placed in a tree instead. Naturalist Hal Harrison, author of a field guide to bird nests, once found a raven nest lined with a thick mat of deer hair, one of nature's finest insulators. I can think of no warmer, cozier way to enter the world than nestled in a bed of deer fur.

Along with pileated woodpeckers, goshawks and other large birds of the virgin forest, the raven suffered disastrous declines during the timber rush of the nineteenth century. By the early 1900s, it had vanished from much of its old range, and biologists were predicting its extinction in many of the remaining areas.

"So completely has this entire region been lumbered over," Pennsylvania ornithologist W. E. Clyde Todd wrote in 1940 about the state's mountainous north, "that I doubt whether even a single raven remains. It is not a bird that can adjust itself to changing conditions, and as the wilderness has retreated, it has steadily retired before the encroachments of civilization ... [W]hether anything can be done to conserve the few remaining individuals is questionable."

But a funny thing happened on the way to extirpation. Ravens did adjust and are currently enjoying something of a boom in the central and southern Appalachians. It may be that they have been aided by the steady maturation of the region's forests, but there seems to be a more general relaxation of the raven's once-rigid standards. Far from requiring large tracts of wilderness, ravens now nest on the sides of active strip mines and quarries, on cliffs overlooking busy highways and within sight of at least one state capitol dome. They are now common even away from the mountains that were once their stronghold, and are becoming routine suburban and even urban birds.

Much the same story can be told of goshawks. Like ravens, they were eliminated as breeding birds in many parts of the Appalachians when the old-growth forests fell, but as the second-growth timber approached maturity in the late years of the twentieth century, the goshawks came back. Shy and secretive for most of the year, goshawks can be ferocious in guarding their nests from threats real and imagined, and the first hint that goshawks have moved back to the neighborhood may be an all out aerial assault. A friend of mine discovered this for himself a few years ago; walking through the May woods one morning, he found himself repeatedly strafed by a screaming female goshawk. He escaped unharmed, but I know other naturalists whose scalps and backs have been ripped open when they strayed too close to a goshawk nest.

The regrowth of the Appalachian forests has helped many bird species recover their former numbers and range, but the composition of those forests has remained basically unchanged for the past century. I have a map tacked to my office wall that shows the major woodland types of North America, a kaleidoscope of color. Pine woods are shown in tones of pink, and these correspond neatly to the southern coastal plain, wrapping west in a long fishhook around the base of the Appalachians. The oak communities (oak-hickory and oak-pine associations, mostly, now that the chestnuts are gone) are assigned green, and these make up the bulk of the southern and central Appalachians, while New England is a patchwork of buff de-

The trunks of paper birch and yellow birch form pale columns in the cool shade of a northern hardwood forest, the iconic plant community of the northern Appalachians.

picting northern hardwood forests of maple, beech and birch. But my eye keeps sliding north to the light purple that signifies the spruce and balsam fir of the boreal forest.

The true boreal forest of the old Canadian Life Zone is a place of pervasive twilight and moist silence, where the thickly clustered spruces lean conspiratorially against one another, closing off the light and the breeze. It is a place to speak in whispers, if at all. There are some hardwoods, but they are scrubby aspens and alders and birches, decidedly second-rate players in a land dominated by three conifers: black spruce, white spruce and balsam fir.

Generally speaking, black spruce prefers wetter feet than the white spruce (hence its old names, swamp or bog spruce), but both are adaptable enough to be found in a variety of habitats. White spruce is usually the taller of the two, although across enormous areas of the North, both species are cut for pulpwood long before they reach their mature height of fifty to seventy-five feet. The white in particular is a well-formed tree, but spruces, even more than most trees, are at the mercy of their environment. A prevailing wind quickly trims all but the leeward branches, resulting in so-called flagged spruces, while on the coast of Newfoundland, the harsh sea winds have sculpted whole groves into bizarre shapes, known by the local name "tuckamore."

Anyone who travels the northern woods comes to recognize and love balsam fir; the pungent scent of its resin is the smell of Canada to many people. "Merely to remember it is to raise before the eyes lake waters, or the soft high swell of the northern Appalachians, or the grandeur of the St. Lawrence Gulf," Donald Culross Peattie wrote. "It brings back the smell of wild raspberries in the sunlit clearing, the piercing sweetness of the white-throated sparrow's song, the bird-like flight of the canoe from the gurgling paddle stroke."

Much of a hardwood forest's beauty rests in its multitudes of green, in the translucence of the color and the play of light and breeze in the leaves. A conifer forest's strengths, on the other hand, are form and texture—the tight spires of the young balsam firs, all bristles and darkly symmetrical, forming perfect, narrow pyramids; the spruces, less manicured and less glossy, the blackish foliage against the gray of the lichened trunks; the tamaracks pale and feathery, drooping their branches absentmindedly about them; white pines at once soft and hard, the blue-green needles cushioning the outlines of the muscular trunks and uplifted branches. Throw in a few birches or poplars to make the wind visible, and you have a forest of nobility and grace.

You do, that is to say, from the outside looking in. It is odd how radically our perception of a forest changes when we are within its arms, and depending on its makeup. The airiness of a broadleaf woodland is in stark contrast to the close-hemmed ranks of spruce and fir. I went bushwhacking once in Baxter State Park in Maine, following a topo map and a compass through the woods. At first the trees were spaced far enough apart to give me a glimpse of what lay ahead, but soon the spruces closed ranks, each one snatching at my face and arms, the branches ghostly with their festoons of old-man's-beard lichen.

The woods were dark and silent; the sound of my footsteps was eaten by the moss underfoot, and there was an unsettling sameness to the view in every direction—360 degrees of uniformity. Progress was painfully slow. For more than two miles, the only change was my periodic passage over small streams and overgrown bogs, where the ground grew unsteady and wet, and my feet sucked and slurped instead of stepping quietly.

Still, I was excited to be making my own trail, watching for moose and fancying that I might even catch a glimpse of a lynx if I were very lucky. I was not aware of the sense of subtle foreboding that was creeping over me—not, that is, until I hit a clearing. Quite without warning, I stepped through the screen of spruces into a small opening, where a tiny pond sat wreathed by sedges, blue against green, with spangles of wild iris like garlands. There was a breeze again, and sun on my face. I lingered there longer than I might otherwise have done, unwilling to go back into the brooding spruces. However much one loves trees—and I love forests above almost all else—the act of walking into a sunlit opening after passing through thick forest unbottles something deep inside, something you might not even have realized until that moment was oppressed. The Europeans who crossed the Appalachians, in the days of the virgin forest, grew starved for sunlight as they walked day after day in the perpetual gloom of the ancient giants. More than once, I've known how they felt.

My wall map of forest types shows the spruce/fir community in patches from southern New England to Maine (I remember the beauty of the mixed conifer and hardwood forest in a New Hampshire autumn, when the sugar maples seem to glow their brightest orange against a backdrop of spruce), then up the northern Appalachians to Newfoundland in an increasingly solid area of purple. But when I look more closely at the map, that same shade of purple runs like a fragmented thread down the Appalachians deep into Dixie, skipping through eastern West Virginia to the corner of North Carolina, Tennessee and Georgia. The map does not discriminate, but these are somewhat different forests from those in the north—red spruce instead of black and white spruces, Fraser fir instead of balsam fir.

Fraser fir is a lovely tree and has the most restricted range of any eastern conifer, found only above five thousand feet in the mountains of southern Virginia, North Carolina and Tennessee. It is the hardiest tree in these mountains as well, surviving higher altitudes (and thus lower mean temperatures) than any other species. Above altitudes of roughly six thousand feet, it is usually the only tree to grow, having left the red spruces to slightly balmier elevations.

It is thought that Fraser firs may have evolved rather recently, perhaps in the five thousand years or so since the last major climatic warm-up, which separated the high peaks of the southern Appalachians from the boreal forests of New England with the wide, hardwood band of the central mountains. The differences between the two species are not great, lying primarily in the bracts between the scales of the cones, which project noticeably in Frasers but not in balsams.

The Fraser fir was discovered by a transplanted Scotsman, John Fraser, who gave up the life of a draper in London to come to the United States in 1784 to pursue his real love, botany. He was traveling with André Michaux, the noted French botanist, who believed that Fraser was trying to steal his hard-won discoveries—a serious concern in the days of early scientific exploration, when reputations (and financial sponsorships) were cemented by bringing exciting new specimens to European attention.

As the story goes, Michaux hatched a scheme to get rid of the Scotsman, claiming that his horses were lost and sending Fraser on ahead while Michaux searched for the missing animals. If this was Michaux's plan, it backfired, for Fraser then discovered this high-altitude fir, which grows abundantly down the spine of the Black Mountains and the Smokies.

The Cherokee were, naturally, familiar with the fir long before Fraser and Michaux stumbled onto the scene; it was commonly used for medicinal purposes, including (according to one source) "such diverse ailments as lung pains, kidney trouble, internal ulcers, colds, venereal diseases, and constipation." The highlanders of the southern Appalachians knew this tree as "she-balsam," because the resin-filled blisters on the bark put the pioneers to mind of milk-filled breasts. (Red spruce, lacking such blisters, became "he-balsam.") Like the Cherokee before them, white settlers gathered the fir resin, poking a knife into each blister and funneling the thick liquid into pots with turkey quills.

First-time visitors to the southern mountains often have a hard time separating the two species, but there is no reason for confusion if the trees are bearing cones—those of the fir stand upright, while the spruce's hang down. Regardless of whether cones are present, one can look at the needles, which are flattened and blunt in the fir but sharp and square in cross-section on the spruce.

When the first whites entered the Appalachians of western North Carolina, the dark, fir-clad summits so impressed them that they named the highest range the Black Mountains. The Blacks and the other high peaks of the southern Appalachians are still dark, but not so much as in the past. Through virtually every fir forest, a gray pall of death has been spreading.

The Blue Ridge Parkway twists down through Virginia and North Carolina from one spectacular overlook to another, passing near the summit of Richland Balsam not far from Great Smoky Mountains National Park. At more than six thousand feet, this is the highest point on the scenic road, and the top of the mountain is just over sixty-four hundred feet above sea level.

Richland Balsam used to be the quintessential Canadian Zone conifer forest thickly grown with mature Fraser firs and red spruces. But on this day, a nasty morning when spring is having a bad sulk and the fog is rolling in, thick and chilling, the summit looks more like a graveyard. As I set off up the trail to the top of the mountain, the fog parts to reveal dead, gray tree trunks strewn haphazardly, overgrown with the arching stalks of brambles, head-high and purple. Other trunks still stand, so weathered that all the color has leached out of the wood, leaving only wraiths behind.

This was a healthy forest until the late 1970s, when a tiny insect known as the balsam woolly adelgid appeared here. A European pest accidentally introduced to North America around 1900, it has caused great damage among balsam firs in Canada but did not appear in the southern Appalachians until the 1950s. Since then the adelgids have spread methodically through virtually every stand of Fraser fir in the mountains, with devastating results. In the Smokies, up to 90 percent of the firs have perished.

An individual adelgid is so small that it looks like a speck of dust or dirt, but its effect on a fir tree is all out of proportion to its size. The adelgid feeds by plunging a specialized, sucking mouthpart into the tree's living outer layer, just beneath the skin of bark. As the insect withdraws water and sap, it injects a chemical, much as a mosquito injects an anticoagulant as it withdraws blood. And, like a human with a mosquito bite, the tree reacts to the adelgid. The chemical injected by the insect causes the tree to produce thickened wood cells that cannot carry sap and nutrients. As countless adelgids feed on a fir, the tree is slowly choked, its conductive tissues sealed off.

The first signs of trouble are yellowish needles that soon become red-brown, a signal that the fir is losing its ability to transport water and nutrients to the crown. On heavily infested trees, the bark and limbs may appear dusted with light snow; the "flakes" are the adelgids, which at maturity secrete a waxy coating from their exoskeletons that protects them from desiccation. By this stage, the fir usually has only a year or two left before it dies.

All along the trail up Richland Balsam, I pass dead firs; the battle here is over, and the adelgids have won—at least with this generation of firs. But at my feet grows a waist-high thicket of young Fraser firs, rising amidst the skeletons of their parents. In many places they crowd so thickly that I cannot see the ground. These are robust, seemingly healthy trees, but the adelgids attack them, too, if at a slower pace than mature trees.

The presence of a replacement generation of firs has given some forestry experts reason for cautious optimism. Unlike the American chestnut, which was wiped out as a reproducing species by the chestnut blight (those saplings that remain keep sprouting from original rootstock), Fraser firs have a chance to reproduce before the adelgids strike. The hope is that each generation will be a little more resistant to the insect's depredations, until a balance is eventually reached like that which exists between the firs and their native enemies. In some high-elevation sites—the Black Mountains among them, along with Mount Rogers in Virginia—some mature Fraser firs already seem to be exhibiting resistance to the adelgids, perhaps because of inherent genetic traits.

The adelgids may not be the firs' only problem, though—or even the most important. It is no secret that the moisture that rains (and fogs, and snows) so abundantly on these mountains is unnaturally acidic. Measurements on Mount Mitchell, where most of the firs have also died, have recorded pH readings as low as 2.12, as acidic as lemon juice, while the average pH of clouds in this region hovers somewhere around 3.7 (by contrast, neutral water has a pH of 7.0). The fogs that made the Smokies and other southern mountains so famous have become toxic enough to injure the trees. Controlled experiments with seedlings showed that the air blowing over Mount Mitchell could slash plant growth by half.

More and more, it appears that air pollution may play an underlying role in the decline of montane Appalachian forests. Lending credence to this theory is the fact that red spruces, which are not affected by balsam woolly adelgids, have also been weakened or killed up and down the Appalachians. In the late 1970s, a researcher in Vermont found that red spruces in the Green Mountains, which he had surveyed in the mid-1960s, were showing across-the-board signs of decline and death. The higher the elevation, generally, the greater the number of dead spruces. He later did core samples (in which a thin rod of wood is removed from a tree, permitting the growth rings to be counted) and found that growth rates took a nose-dive after about 1960, years before the damage became visible. Most frighteningly, the results were much the same from New Hampshire all the way to North Carolina.

Foresters working in the southern Appalachians have found the same evidence of reduced growth rates in Fraser firs, too. In fact, you needn't be a scientist to see this for yourself. Climbing the trail up Richland Balsam, I paused here and there where dead trees had fallen across the path and had been chain-sawed in half. It takes just a minute to count the growth rings, and after I checked more than three dozen trees, a clear trend emerged. One fir was typical. For the first thirty-one years its growth was robust, with plenty of space between the rings. Then there was a dramatic slowdown for the next fifteen years, with the rings crowding each other. Then death. The story was much the same with the other dead firs I checked—healthy growth for twenty or thirty years, then a swift decline into death. Annual growth in red spruces on Mount Mitchell, fifty miles to the northeast, has declined more than 80 percent since 1960.

Almost all trees slow their growth as they approach maturity, of course, but this is something different. All across the southern Appalachians, Fraser fir and red spruce showed a significant decline in growth rates about fifteen years before the mass deaths began. One theory is that air pollution, including acid deposition and ground-level ozone, seriously weakened the trees, opening the door to disease and pests. If true, the adelgids were only the final, fatal insult.

Sadly, the Fraser firs were only a precursor to a much more widespread—and damaging—adelgid threat. Throughout the central and southern Appalachians, you can find hemlocks that look as though they has been lightly sprayed with the fake snow people use at the holidays, a chunky white flecking all over the lower branches and twigs. It is a depressingly common sight—the waxy covering of a tiny insect called the hemlock woolly adelgid, a pest that has spread with terrible speed.

Hemlock adelgids are minute, and millions may infest a single tree, sapping its energy and killing it, much as the balsam adelgid kills Fraser firs. Apparently accidentally introduced from Asia to the West Coast in the 1920s, the insects hit western hemlocks, which showed a fair resistance to their attacks. By the 1950s they had appeared in the mid-Atlantic region, possibly as a result of infected nursery stock, and began to target eastern hemlocks.

Unfortunately, eastern hemlocks lack their western relatives' tolerance for the adelgid, and the pest has spread across a huge swath of the hemlock's range, from southern Maine and the lower tier of New York

Stark as a graveyard, the dead trunks of Fraser firs still stand on the summit of Mt. Mitchell, where an imported insect pest killed almost all of the mature trees.

south to Georgia and west through the Appalachian Plateau country. Millions of hemlocks have died, and millions more barely cling to life, their lower branches dead, and thin clusters of sickened needles clinging to the outermost twigs, sparse and meager.

The loss is devastating on several levels. The most obvious is the tree itself, one of the most graceful of eastern conifers, providing a habitat anchor for plant species that require acidic shade. Its dense foliage provides nesting sites for everything from warblers to sharp-shinned hawks. Hemlocks thrive best in cool, damp locations, especially on north-facing slopes and along streams, and in many parts of the Appalachians they provided a solid canopy over headwater creeks, keeping the water cold through the hottest days of summer.

Stripped of their protective cover, the streams are left exposed to extremes of water temperatures—summer peaks too high for many forms of cold-water aquatic life to tolerate, and colder-than-normal temperatures in winter. Brook trout are only the most visible of the victims, which range from other fish (including some unique to the Appalachians) to a host of invertebrates. And ecosystems being the intricately knotted units that they are, the ripples spread far beyond the dead hemlock groves and overheated creeks, with effects throughout the forest at which we can only guess.

There is little that can be done to control the adelgids artificially, beyond the most immediately local basis. Trees can be sprayed with insecticidal soap, and the soil can be drenched with chemicals that the trees take up, killing the pest. While this may work for a few specimen trees in a backyard, there is no comparable treatment that works at a landscape level. (That said, Shenandoah National Park, which has lost almost all of its hemlocks, is fighting a last-ditch attempt with systemic insecticides and sprays to save a handful of mature trees—a gene pool for the future, should the adelgid epidemic eventually be stemmed.)

So far, the infestation has not expanded into New England, because the adelgid has one weakness: cold. Temperatures below minus four or five degrees Fahrenheit are usually lethal to it—but such cold spells are increasingly scarce, as the climate warms. The rare frigid blast that plunges the East into a deep freeze is cause for celebration among those battling the adelgid. In the Great Smokey Mountains, where the pest arrived in 2002 and has killed many of the park's ninety thousand acres of hemlocks, mountaintop temperatures as low as minus twelve in 2014 provided at least a temporary respite.

The only serious hope is a functional biological control, but that has proven elusive. Several species of Asian beetles have been carefully tested and released, but the results have been lackluster at best. The current hope is a predatory beetle from the Pacific Northwest, which may be the reason western hemlocks have been able to weather the adelgid with little trouble. But across thousands of square miles of Appalachian forest, the worst damage is already done. The hemlock glens that were once bathed in chill, green light lie open to the sun, the trunks of the dead standing among the withered, gray branches of the almost dead.

A year or so after the first adelgids appeared in the mountains around my house I bushwhacked, with a great deal of trepidation, along the top of the Kittatinny Ridge. Leaving the trail, I followed the cracked boulder fields for a mile or so, to visit the finest hemlock I know.

It was not a giant tree like the old-growth hemlocks of the Smokies, but it was a monarch. It grew from a deep crevice in the boulders, rising only about forty feet but with a tremendous spread and a thick, divided trunk; size notwithstanding, this was a very old tree, probably overlooked by timber cutters in ages past because of its inaccessible location and odd shape. The lower boughs cascaded out and down, then swept up at the tips, curtaining the gray rocks with smoky, blue-green needles. Around it, wind-hardened chestnut oaks and sweet birches grew, a court for royalty, and the orange berries of mountain ashes decorated the forest like a tapestry.

In one of the deep recesses beneath its branches, down among the jumbled rocks, I smelled the sharp scent of cat and found a worn, flattened place beneath a sandstone ledge where, I guessed, a bobcat came to sleep. Sadly, I also found adelgids—not many, but enough. I sat for a long time beside the old hemlock, listening to the wind crying in its needles, trying not to think of this as a deathwatch.[1]

In the course of the Appalachians' long, long history, the mountains have seen many species come and go, like brief lights that burned and flickered out. Much of the chain's history is so ancient that it, too, has been erased by time, but the ice age is so recent (indeed, scientists like to point out, were it not for global warming we would still be in the ice age, merely enjoying a brief surge of mild climate between continental glaciers) that its signs are easy for anyone to read. In dribs and drabs down the Appalachians, the work of beavers, ice or the quirks of geography, exist hundreds of such Pleistocene echoes, the relict bogs.

Bogs are common in the northern Appalachians—in fact, peatlands (which include bogs, fens and related wetlands featuring accumulations of sphagnum moss) are among the most widespread of the nonforest habitats from Maine through Newfoundland. But south of Maine, bogs are much rarer, sheltering plant and animal refugees from the days of the last glacial retreat, slices of the North embedded in the South.

There is no dry way to enjoy a wild bog. By the time you cross from the edge of the upland forest and pass under the drooping boughs of the black spruces, the ground has become wet; rivulets collect into streams, framed by stark green moss and the tiny leaves of goldthread, and every step you make sinks a little deeper into the peaty soil. At first, you are tempted to hop from one drier hummock to another, but sooner or later a slick mound turns under your ankle and you plunge knee-deep into the brown slurry of peat and cold water that hides just beneath the layer of sphagnum moss, and you give up any hope of staying dry. Hip boots help, but on a hot summer day this dunking is really quite pleasant and is enough to make you forget about the cloud of mosquitoes and gnats that halo your head.

Most northern bogs are direct descendants of the glaciers, often formed when the retreating sheet left a chunk of ice embedded in the soil (if this seems improbable, consider for a moment the immense weight of a glacier more than a mile thick). When the ice melted, it left a lake behind as evidence of its passing. Gradually, with the passage of thousands of years, the surrounding forest reasserted itself, as water-loving plants—especially sphagnum moss—encroached around the edges to form floating gardens. The mats expanded and thickened, providing a steadily drier, more stable base for less water-tolerant species like black spruce, while century after century, the dead, partially rotted sphagnum accumulated as a thickening layer of peat.

Eventually, the cap of sphagnum meets in the middle and seals off the open water for good, but beneath its thin, growing layer lies a lens of water and peat upon which the whole mass still floats. Slogging across one of these "quaking bogs" is like traversing a huge water bed, into which you sink up to the shins and occasionally the hips. It is an unsettling experience to step onto what looks like solid ground only to have it give beneath your feet and ripple at each footstep, or to watch the trees sway drunkenly at your approach.

Not every bog started as an embedded ice chunk. Some were formed when rockslides plugged up creek drainages, creating stagnant backwaters, others by massive pieces of dirty, sediment-filled glacial ice left behind as the sheets retreated. As the ice melted, the rocks and mud fell free, forming roughly circular dams of heavy material, lined with finely ground particles that acted as a water seal. (You can see this same principle at work on a smaller scale by dumping a big scoop of gravel-filled snow in the sun. As the slush melts, the gravel accumulates on the outer rim, leaving a ring of pebbles behind when the snow is gone.)

The most recent continental ice sheet made it only as far south as the Poconos, yet bogs are found hundreds of miles farther south, particularly in the highlands of West Virginia and in smaller numbers through North Carolina. Obviously, these were not created by glacial dams or embedded ice (poor drainage was most often the cause), yet they, too, are ice age remnants, providing refuge for northern species when the climate turned to warmth. Cranberry Glades in West Virginia, the state's largest bog complex and now a federally protected botanical area, is the southernmost range limit of a number of animals and plants, among them hermit and Swainson's thrushes, northern waterthrush, mourning warbler, purple finch, buckbean and bog rosemary.

The cold, damp forests that rim the bogs have their own specialties. If the ground is rocky and the forest a mix of hardwoods and conifers, the area may shelter a colony of long-tailed shrews, found only in the Appalachians and the Adirondacks. Rare and little-studied, these five-inch insectivores forage in tunnels dug beneath mossy rocks and boulder piles, often along the margins of tiny streams. The species is found in a narrow band in the mountains from New Brunswick to the Smokies, and aside from a few educated guesses about its breeding cycle and diet, almost nothing else is known about it.

Mammals have more mobility than, say, salamanders, but it has always struck me as odd that the Appalachians don't have more endemic species of mammals—varieties found nowhere else but in the mountains. Besides the long-tailed shrew, the only other example is the Appalachian cottontail, a rabbit found as far south as Georgia and Alabama and up the mountains to Pennsylvania and New York.

Virtually indistinguishable from the eastern cottontail of backyards and farms, the Appalachian cottontail is smaller and darker, and it almost always has a black spot between its marginally shorter ears, a trait present in only some eastern cottontails. Eastern often have a white blaze on the forehead, something never seen in the Appalachian animals. Further confusing matters is the more northerly New England cottontail, with which the Appalachian form was lumped until 1992; they are both so similar to eastern cottontails that no one realized there was more than one species until 1895.[2] Appalachian cottontails are creatures of dense heath stands and conifers, like the blueberry/laurel barrens and red spruce forests of Dolly Sods in West Virginia, although they may have once been more common at lower elevations.

Hunters had long differentiated between lowland and highland cottontails, though. They knew the Appalachian cottontail as the "woods rabbit" or "mountain rabbit," for it requires dense forests with a thick understory of shrubs and herbs, rather than the broken, brushy habitats favoured by eastern cottontails. The woods rabbit is the game of choice for many beagles, because it is reputed to run better before the hounds, leading them on exciting, hours-long chases instead of holing up in the first available woodchuck den as farm rabbits tend to do.

Predictably, the clearing of the mountain forests was a blow to Appalachian cottontails, for their preferred habitat was supplanted by the kind of brush favoured by eastern cottontails. Making the problem worse, for most of the first half of this century, state wildlife departments and private hunting clubs imported huge numbers of cottontails from the Midwest for release in the East, creating a mongrel rabbit of mixed genetic background— always a formidable opponent in the tug-of-war of natural competition. Although that practice has stopped, in some areas Appalachian cottontails were simply swamped, so that today their range through the central and southern Appalachians is a series of pockets, some large and some small. Fortunately, biologists now recognize the importance of preserving unique species like the Appalachian cottontail and are managing some mountain habitat to its benefit.

As mentioned earlier, not every bog is the direct progeny of a glacier. One of the finest relict bogs in the central Appalachians hides at the end of a long, dusty dirt road about five miles from State College, Pennsylvania. Known as Bear Meadows, it is unusual in that it lies within the ridge and valley province, a part of the Appalachians spared by the last glaciers, and thus with few natural lakes or bogs. Dated core samples from its bottom, analysed for fossil pollen, suggest that Bear Meadows has been a bog for nearly ten thousand years, probably due to repeated damming of Sinking Creek, which flows through it, by colonies of beavers.

But beavers are common in the Pennsylvania mountains, while black spruce bogs are not. What makes Bear Meadows unique is an accident of topography. The bog occupies a shallow bowl between Tussey and Thickhead mountains and acts as a sink for cold air generated by the higher hills, which then slides down the slope and collects in the flattened valley; this action is reinforced by the bog's altitude of nearly two thousand feet, quite high for Pennsylvania. Most other bogs in the central and southern Appalachians, including Cranberry Glades, rely on this same combination of a sluggish water source and an unusually cold local climate. The chill air and wet, acidic soil provide a fragment of the conditions that existed when the bog was formed millennia ago, and a haven for the Canadian Zone plants that once lived widely across the mid-Atlantic states.

At Bear Meadows, the black spruces and balsam firs grow thickest around the northern edge of the five-hundred-acre bog. The forest here is a confused tangle of dead branches and jabbing twigs, interlaced so you must bull through them with graceless force, snapping limbs like an ox. At length, and after more than a few scratches, the spruces give way to the open bog, with the thin blue line of the creek running down the middle, framed by stands of green sedges. The ground in the bog is a shaky mat of sphagnum and, below that, peat, more water than soil.

It is impossible to imagine a bog without sphagnum moss; it would be like trying to imagine a forest without trees. Sphagnum mosses—there are dozens of species—quite literally create the bog from what was once open water, and they provide the foundation for most of the other plants that inhabit it. Buoyant and amazingly absorbent, sphagnum can hold between fifteen and twenty-five times its weight in water—as anyone who has tried to moisten peat moss (the dead, dried remains of sphagnum) can attest. The dried plants were used by Native Americans as diaper padding, and their antiseptic properties have made them a valuable wound dressing all over the world.

In the full sun beyond the verge of the conifers, a thriving plant community lives on and around the moss foundation: highbush blueberries hung indigo with fruit in July; spiraea with its feathery, purple plumes of flowers; cotton grass, leatherleaf, wild viburnums. Each species must contend with the difficulties of growing in a medium that is almost devoid of nutrition, for in the bog's icy, acid water very little rots, and so very little can be recycled to the clustering, impatient roots. In fact, the acidity and the pervasive cold interfere with a plant's ability to absorb water, so that many species have evolved moisture-conserving features like leaves that are thick and leathery, have rolled-under edges or undersides that are densely furred. These are many of the same adaptations seen on alpine tundra plants, and for the same purpose.

Orchids, which lack most of these adaptations, seem to deal quite well with the bog water's acidity; a single bog may host a dozen or more species, some stunningly beautiful. I clearly recall the first time I came across a trio of *Arethusa*, popularly known as dragon's-mouth orchids, rising from the sphagnum of a bog in Maine. The pink flower defies easy description—an open, down-curving "mouth," lined with yellow hairs and rimmed with a ruffled lip, and three upright sepals. The three orchids glowed in the sun,

The boreal forest runs like a unifying thread down the spine of the Appalachians, from Canada and northern Maine, where dense stands of white spruce and balsam fir dominate, to the highest elevations of the southern mountains, where red spruce and Fraser fir hold sway.

and I stood gaping for so long that I sank to my shins in the moss. Orchids do so well in because they have allies—mycorrhizal fungi that live within their roots in a symbiotic relationship and actually perform the task of eking out a living from the acidic substrate.

A number of plants have neatly sidestepped the bog's nutrient deficiency by becoming carnivores. They are not monster movie man-eaters, needless to say; the largest is only as big as a human hand, and their prey are insects. The most common is the round-leafed sundew, a dainty plant with a rosette of paddle-shaped leaves, each about the size of a thumbtack, that grows among the clumps of sphagnum moss. The sundew's leaves glisten, for each is covered with a star burst of tiny, glandular hairs gobbed up with a sweet, sticky fluid secreted by the plant—a fluid attractive to flies and other small insects. Once in contact with the hairs, however, the insects cannot free themselves, and their struggles only cause the leaf paddles to fold, so that more of the hairs arch over and entomb the bug fully.

Once dead, the insect is dissolved—"digested" is not too precise a word—by enzymes released by the sundew. Of particular importance to the plant are the nitrogenous compounds in the dead insect's tissues, since the acid bogs are notoriously nitrogen-poor, and this element is crucial to plant development.

Other carnivorous plants take different approaches to hunting. The pitcher-plant (now quite rare in Bear Meadows due to collecting for the plant trade) grows red-veined leaves that form upward-curving tubes filled with rainwater and surrounded flaring lips. Viewed closely, the insides of the tubes are covered with minute hairs, which point down—but fortunately for the pitcher-plant, insects never look closely. Lured by the water and the faint, cloyingly sweet scent produced by the plant, a beetle or an ant tumbles into the trap. Being buoyant, it should float, but the pitcher-plant's secretions weaken the surface tension of the water, and the insect sinks like a stone.

Scrambling to the tube wall, the insect tries to climb out, but the carpet of sharp bristles defeats it. At length it drowns, and within its chitonous exoskeleton, the plant's enzymes turn softer tissues to soup, which leaches into the water and thence through the porous walls of the pitcher—a dose of badly needed sustenance in a miserly environment.

As I stand submerged up to my knees in the Bear Meadows bog, the frigid water has seeped through my old sneakers and wool socks and begins to rob my toes of feeling. A tiny hoverfly drones past, circles once and drops toward a sundew, which has raised a thin, curved wand of pink flowers to the sun. The fly feeds at the single open blossom, then buzzes down to the leaves. It brushes a hair, becomes snarled in the glue for an instant, then breaks free and flies, somewhat unsteadily, out of my view. For the sundew, the bog's poor meal will have to do for another day.

The massed fist of the southern Appalachians spills over from the Smokies into Georgia, an expanse of buckled terrain that covers the state's northern tier. The mountains are not quite as high as those to the north, and winter is evicted here more easily. Thus, when a late-April snowstorm closed the Smokies one recent spring, I turned gratefully south into Georgia.

I wound down endless curves and switchbacks from the mountains to the valleys and back up again, the sort of torture that Appalachian roads routinely inflict on drivers in a hurry—but I was in no hurry at all, and the twisting way was a delight, especially with an extravagance of dogwood lighting the undersides of the corridors of graceful tuliptrees. In the sheltered valleys, plowed fields married strips of rich, red soil with the vivid green of newborn spring foliage. The land was a surfeit of beauty and fertility.

In the lowlands between the hills, the forest was laden with white oak; on these two-lane back roads, you'll find hand-lettered signs advertising white oak baskets, a specialty of the South. Down most of the Appalachians, white pine is the most common conifer in such upland habitats, but here shortleaf pine appears with increasing frequency among the oaks, shading an understory of holly, dogwood and wild azalea. Shortleaf isn't as tall a pine as the white, and its importance as a timber tree ensures that few will live to anything approaching old age; indeed, I passed lumber trucks all day as I ambled, many of them carrying the distinctive, orange-barked trunks of shortleafs.

Dusk found me at Water's Creek, a small campground in the Chattahoochee National Forest. The evening was still and almost silent, bereft of birdsong; in fact, the only thing moving besides me and two other campers was a Carolina chickadee excavating a nest cavity in a dead tree across the stream from my tent. Every minute or so it poked its masked head out of the hole, fibers of rotted wood sticking out either side of its bill like a stiff, pale mustache. Then it shook its head violently, like a sneeze, scattering the punky wood on the water below.

Morning was as busy as the evening was quiet. Flocks of migrating songbirds, just finished with a night of travel, had pitched into the forest to feed and rest, and the dirt road that led uphill from camp was crawling with birds. Spring was working on them, for the males were singing incessantly, still hundreds of miles from their breeding grounds. Within half an hour I found eleven species of warblers, including a dazzling male Cape May warbler, several bay-breasteds and a worm-eating warbler—this last patiently working his way up through a tangle of honeysuckle, prying into clusters of dead leaves.

The worm-eating warbler's habit of exploring suspended leaf clusters from the bottom up has been noted by ornithologists working in the Caribbean and Central America, where this species winters, and is one of the bits of evidence that exploded the comforting "adaptable migrant" theory once in vogue. Biologists originally thought that North American migrants like warblers, vireos and thrushes would be more adaptable, in the face of tropical deforestation, than the resident species. The idea seemed plausible; the migrants could well be the perennial new kids on the block, squeezing in wherever there were openings left by the permanent inhabitants of the tropical forests. If so, they might fare better as the virgin rain forests were converted to scrub or agricultural land.

No such luck. Researchers have shown that migrants from North America have their own, highly refined niches in the tropics. On the northern breeding grounds, worm-eating warblers scour leaves of any sort for insects, but once in the tropics for the winter, they concentrate almost exclusively on dead leaves in suspended tangles—a behavioral adaptation that presumably minimizes competition with other migrant and resident birds. Nor, as it turns out, is such specialization unusual; sexes within the same species may even have radically different (and equally rigid) habitat requirements on the wintering grounds.

After my morning of bird-watching, I spent the rest of the day poking around back roads and through quiet hollows, so that by late afternoon, as a result of serendipity rather than planning, I found myself in the mountains south of Blairsville, heading for Brasstown Bald, at roughly forty-eight hundred feet the highest point in Georgia. I passed the last tourist of the day heading down the mountain as I drove up, my car laboring against the steep, winding ascent. So many of the tallest Appalachians in each state—Cheaha Mountain in Alabama, Mount Mitchell in North Carolina, Mount Washington in New Hampshire—have roads right to the top, accommodating the inactive multitudes. The road up Brasstown Bald is gated a half mile short of the summit, but during the tourist season you may purchase a bus ride to the top to save yourself a hard walk.

The grassy bald, as such, is gone; the top of the mountain now sprouts a big U.S. Forest Service visitors center, handsomely done in wood siding but hardly a fitting substitute for a wild summit. Everything was closed for the day, the interpretive displays locked, so I walked around the circular observation deck, staring off into four states—the crumpled Smokies to the north, Tennessee to the northwest, the hills of Georgia in every direction. Their names hinted at history: Blood Mountain and Slaughter Mountain, Tesnatee Gap, Sosebee Cove, Hogpen Gap, Wolf Pen Gap.

The forest that covers Brasstown Bald's upper reaches is composed mostly of gnarled, wind-sculpted white oaks, with hemlocks scattered here and there and a thick understory of heaths. It is very much a southern community. Only right at the summit, relegated to the fringe of the lawn and the road around the visitors center, are there a few spruces, starchily erect in the evening breeze.

Yet even here, so far into Dixie, the ravens keep faith with the ancient Canadian legacy. A pair rode the updrafts, holding almost motionless in the upwelling of wind that deflected against the mountain, just at eye level with me and a long stone's toss from the edge of the hill. The angle of the late-day sun was such that the ravens did not appear black but almost sparkled with a vivid purple iridescence that shimmered over their heads and across the upper surfaces of their shoulders and flight feathers. Spring was working on the ravens, too. The nearest pulled in its wings partway and slid forward at a steep angle in a courtship display, rolling onto its back as it dove. It croaked twice, then flared its wings, rising like a pendulum against the setting sun.

Their presence, and that of the tattered spruces, moved me profoundly. Here, among the diminishing mountains that rise not quite high enough, the Canadian Life Zone finally loses its fight with altitude, latitude and temperature. Here the plants and animals of the North were sheltered from the glaciers, biding their time for a warmer day, but now the times have grown too warm, and the island refuge has been swallowed—all but these last few castaways. On this Georgia hill, as the sun turned the oaks gold, there was only the croak of the raven to sing requiem for what once was.

Bull Elk

Chapter 6

THUNDER, DIMLY HEARD

November 1867, and a snowstorm was blowing through the flat-topped Alleghenies of north-central Pennsylvania, plastering the woods with wet, sloppy flakes. Jim Jacobs, a full-blooded Indian from the Cattaraugus reservation in western New York, slogged through the spruce bog known as Flag Swamp, watching the ground. The snow was filling them fast, but he could make out the deep hoofprints of an elk, nearly five inches across, strung out in a swaggering trail through the woods. A bull, and a big one. Jacobs cocked an eye at the sky, feeling the deepening storm, but slipped the leads off his hounds anyway and set them on the trail. The dogs snuffled the snow a few times, then lit out, baying deeply, as Jacobs followed.

Soon the trail's character changed; instead of the evenly spaced prints of an animal walking slowly, it showed wide gaps where the elk had broken into a gallop, aware that he was being followed. Jacobs now moved at a run himself, slipping periodically on the treacherous, sodden ground.

No one knows exactly how long Jacobs trailed the elk through the early winter storm, or precisely where he went. Some say the chase ended near the source of the Bennetts Branch of the Sinnemahoning, others that it was across the divide on the headwaters of the Clarion River that the great elk turned to make his stand—for a big bull would always come to bay eventually, the old-timers swore, usually picking a jutting rock ledge where he could turn his back to the safety of open air, lower his massive, ivory-tipped rack, as thick at the base as a strong man's forearms, and gore bloody hell out of anything that dared to come within reach.

Such a strategy had worked against wolves for thousands of years, but not against men with guns, and not this time. One can imagine the scene: The elk on his rock, mouth open and ribs heaving; Jacobs' pack of hounds hemming him in; the flat light of a snowy day with the forest fading to white; the man catching his breath, then leveling the rifle and shooting.

In so doing, he killed the last native elk in the Appalachians—and in all likelihood, the last of its breed anywhere east of the Mississippi.

That the elk, the *wapiti* of the Shawnee, should come to such straits still seems difficult to believe. This was once the most widespread game animal in North America, from western Vermont and Massachusetts down through New York, Pennsylvania, Maryland and in the mountains to Georgia, then west across the Plains to the Pacific. During the first years of settlement, elk were apparently found clear to the Atlantic seaboard; records from the 1650s speak of them being commonly shot just outside of what is now Philadelphia. They vanished quickly from the lowlands, however, and by the middle of the eighteenth century the center of their abundance in the East was in the hills of Virginia and central Pennsylvania, where they congregated along stream valleys and near natural saline licks.

The next century was no kinder to the wapiti. By the 1790s, settlers were fanning out along the Susquehanna's west branch, probing up the larger tributaries for homesteads. They (and the logging crews that followed) needed meat, and the elk was once the largest and most convenient source. One such family was the Tomes, who in 1791 purchased land near the mouth of Pine Creek, just west of present-day Williamsport, and proceeded to make quite a name for themselves as elk hunters.

Jacob Tome and his boys were the best—at least in the opinion of Philip Tome, one of those sons, whose *Pioneer Life, Or, Thirty Years a Hunter* is a lively (not to say imaginative) account of meat hunting in the Pine Creek region during the early 1800s. Tome writes of hunting elk by firelight while poling down the river on rafts, of contesting with wolves for his kills, but none of his tales compare with a bizarre episode in which his father won a bet by capturing an Appalachian bull elk alive.

The wager, for the smacking sum of 250 pounds, was made late in 1799 between Tome and the local tavern keeper, who gave Tome four months to bring in a live elk at least fourteen hands tall—no half-grown calves would satisfy the bet. Once tracking snow came in January 1800, Tome, Philip, another son and a fourth man set out with five hounds, including a trained elk dog on loan from a neighbor.

After two days of searching, Philip's brother found elk sign, and they broke camp and began to follow them. Over the next four days, the party followed a large bull across a zig-zagging course that, according to Tome's reckoning, covered more than sixty miles through these lung-busting mountains. Their dogs brought the elk to bay several times, only to lose him. Finally on the fifth day, having swung back toward the Pine, the elk picked a rock and made a stand.

"At 8 o'clock [in the morning] we began to maneuver," Tome wrote. "We tried first to throw the rope over his head, but he jumped from the rock and broke away. We then let all our dogs after him ... He ran about half a mile but the dogs pursued him so closely, and closed in on him so often, that he wheeled about and returned to the rock."

With the dogs to divert the elk's attention, the elder Tome slipped around to the upper side of the ledge, rigged a noose on a twenty-foot sapling, and dropped the rope over the elk's antlers. More ropes snubbed him to a tree, and with a horse they managed to get him back to the settlement.

"This was the first elk caught alive full-grown along the waters of the Susquehanna," Tome concluded. "He was 16 hands high and his antlers were five and a half feet long, with 11 branches."

Through the early 1800s the pressure on eastern elk grew steadily, and the end, when it came, was swift. The winter of 1852 was the last that elk "yarded up" in the Alleghenies, a pitiful remnant of twelve on the Clarion River, seven of which were killed by trappers. During the 1860s there were more rumors than elk in the woods, and until Jim Jacobs' hunting trip in 1867, no one had seen a live wapiti in several years.

Little is left now but the names: Elk City, Elk Garden, Elkhorn, Elk Run and Elk Creek. There's an Elk County in Pennsylvania, encompassing the mountains near the Sinnemahoning where the last bull was run to ground. But in one of those exquisite ironies that history likes to play, there are today elk along Bennetts Branch, presumably not far from where Jacobs fired *finis* so long ago—hundreds of them, scattered in small bands through the second-growth forest of oak, beech and hemlock.

These are eastern elk by conveyance, not blood, the descendants of more than one hundred Rocky Mountain elk trucked in from Yellowstone between 1913 and 1926 and released in the Appalachian Plateau country of northern Pennsylvania. (Others were released to the east, in the hills south of the Poconos and near State College, but they soon died out again.) The external differences between the native elk and these immigrants are slight, judging from old museum specimens: marginally smaller antlers in the newcomers, and possibly smaller in body size, too, if the reports by nineteenth century hunters of thousand-pound bulls can be believed. Presumably, though, eastern elk were better suited to forest life than the open-country, grazing western race. The eastern elk was once considered a separate species, then downgraded to a subspecies, *Cervus canadensis canadensis*. Now the American elk itself has been lumped with the red deer of Europe as *Cer-*

vus elaphus, and to a latter-day woodsman wandering the Bennetts Branch valley, it is merely enough that there are wapiti in the hills.[1]

It is a mild spring day, when the oak leaves have that vibrant, almost neon green that lasts for only a week or two before the deeper, bluer green of summer sets in. Foamflower is blooming along Hick's Run, a stunningly beautiful trout stream that flows through a steeply cleft valley north of the Sinnemahoning; fuzzy leaves of pink lady's-slipper orchids peek through the dead oak and beech leaves; stands of mayapples and trout lilies cover the damper patches near the creek. Flashing black and white, a pileated woodpecker swoops across the creek, an exclamation point of crimson from his crest, while high overhead I hear the piercing skirl of a broad-winged hawk newly returned from the tropics.

As quietly as I can, I work my way upstream, watching for elk sign. This is not the best time of the year to be looking; autumn, when the bulls are bugling out challenges to one another, or a winter's day with a good tracking snow, would be better choices. But I was passing through and had some spare time, and I thought I'd make a detour pilgrimage in honor of the old bulls and the men that hunted them.

There are plenty of deer tracks in the mud along the creek and scatterings of their pellet droppings, but I find none of the cowlike tracks of elk or their big, oblong scat. After a fruitless hour I veer away from the stream and up a side gully, pushing to the flat bench on top, where I settle down with my back to a tree and catch my breath, thinking wryly about Tome and his sixty-mile pursuit; an elk would have nothing to fear from my feeble chase. The day has gone warm, and the sun on my face feels fine. The dirt road where I've left the car is a good way off, the paved road considerably farther, so there's no sound in the woods but the wheedling of a vireo.

Eyes closed, I try to imagine how this land looked in Tome's day. This was Pennsylvania's Black Forest then, rich in virgin white pine and hemlock; rich, too, in wildlife—herds of elk along the rivers, packs of wolves to hunt them, solitary mountain lions, bobcat and lynx, fisher, marten, snowshoe hares, at least a few wolverines. This was very much a northern world, and the list of animals it supported reads like a ledger of the lost.

The Appalachians of today, bountiful as they seem to modern eyes, are a paupered ecosystem. Pennsylvania, still known for the abundance of its wildlife, has since colonization lost nine species of large mammals and furbearers, among them bison and native elk, along with several birds, including the passenger pigeon, once the most populous land bird in the world.

Predictably, the carnivores have suffered the worst; while game animals like elk were shot out because of greed, predators like wolves, lynx, moun-

Reduced to shy flocks in the most remote part of the mountains in the early 20th century, wild turkeys have reclaimed virtually the whole length of the Appalachians, including suburban backyards such as the one where this hen and her poults forage in a flower bed.

tain lions and bears were the victims of zealously applied hatred. Bounties for wolves were in place as early as the 1600s in New England and among the Swedish settlers along the Delaware River, and by 1682, Pennsylvania's provincial government was paying ten shillings for a male, fifteen for a bitch. More than two centuries later, in 1897, New York was still paying a handful of wolf bounties.

Bounty records provide a depressing snapshot of the former abundance and swift decline of the great predators. In just twelve years, from 1808 to 1820, 562 wolf scalps were turned in for bounty in Luzerne County, Pennsylvania—273 in one year, perhaps the result of organized "wolf drives" in which every able-bodied man in the community beat the woods over huge areas, forcing the local wildlife past waiting marksmen.

The ire of the settlers was not unfounded, for wolves and mountain lions did take livestock—a heavy loss to a family with a single milch cow or draft horse. But much of what motivated them was simple fear, reinforced by centuries of experience with the more dangerous and aggressive European wolf, which, unlike its North American counterpart, had a history of killing humans.

The last wolf in the Pennsylvania Appalachians was probably shot in 1892 in Clearfield County, a little south of where I now sit in the springtime sun, although some people claim the species grimly hung on until 1901, just nudging under the wire of the twentieth century before being snuffed out. Mountain lions were under the gun as well; one central Pennsylvania county paid out nearly six hundred bounties during the twenty-five years ending in 1845, sixty-four of the payments to just one hunter.

The story was the same everywhere. Wolves were gone by 1860 in Maine and by 1905 in North Carolina, and the Adirondacks in New York provided a refuge only until the turn of that century. Mountain lions, always harder to get a grip on than the noisy, social wolves, faded out around the same time north of Virginia (Vermont's last was killed in 1881), although a few survived the first half of the twentieth century in the southern Appalachians.

Pennsylvania's Black Forest gave up its last wolverine in 1858, its last fisher in 1901, its last marten around 1900. The only fully documented lynx there was killed on November 10, 1923, so incredibly late a date that the animal's origins must be suspect—probably an escapee or a wanderer from Canada. Considering how logged clean the hills were, it seems impossible that it was a native survivor.

Of course, the fauna of the Appalachians was in flux even before the white settlers appeared with gun and ax. For several thousand years, based on samples of fossil pollen taken from bogs, the climate had been warming in eastern North America. The last, lingering ice age mammals, most notably the American mastodon, faded to extinction about nine thousand years ago, no doubt aided in their departure by Paleoindians, who hunted them with sophisticated and powerful spear-throwers.

The cool spruce forests inched north, taking many of their endemic species with them. A moose antler turned up at a salt lick in the Alleghenies, the only real evidence that the species occurred south of the New York line in near-historical times.

Only in the most northerly reaches of the Appalachians did vestiges of the ice age live on. Woodland caribou, the largest subspecies of reindeer in the world, were once found in Maine and perhaps northern Vermont and New Hampshire; one was even recorded in western Connecticut. Logging in the 1800s destroyed their habitat and made them vulnerable to shooting, but it had a more perfidious effect—it opened the North Woods to white-tailed deer, which brought with them a parasitic brainworm that, while apparently harmless to whitetails, is deadly to caribou and moose, causing lethargy, a sort of premature senility and death. By 1908 the caribou were gone from Maine, although herds still exist in the Canadian Appalachians, including the Shickshock Mountains and Newfoundland.

Most of the wildlife of the primeval Appalachians is well documented. While questions remain (were red foxes, for instance, native to the central mountains, or are they descended from European captives released for fox hunting in the eighteenth century?), the early explorers and naturalists left us a fairly complete outline of distributions, relative abundance and physical descriptions. Except, that is, for the bison. This largest of the Appalachians' native mammals is also by far the least known.

Some authorities question whether the bison ever existed in the mountains as anything but a periodic stray from the prairies to the west, while others (with justification) cite the wealth of anecdotal evidence that suggests buffalo herds were common as far east as the Susquehanna River. In 1876, a skeptical J. A. Allen wrote: "The Alleghenies may be taken as its general eastern limit, its occurrence in the mountainous and elevated parts of the Carolinas being due rather to the occasional wandering of small bands through the mountains from the immense herds that formerly inhabited the valleys of West Virginia and the adjacent parts of Kentucky and Tennessee ... I have failed to find a single mention of the occurrence of this animal within the present limits of New York, New England [or] Canada ... that will bear a critical examination."

Some authors, citing texts from the late seventeenth and early eighteenth centuries that mention buffes, buffles, wild bulls, wild cows and wild cattle, have extended the bison's range right up to the Atlantic and into Canada. Allen noted that the records could just as easily be referring to elk and (in New England) moose—and that if these are accepted at face value, then so too must claims of monkeys and apes in Virginia and wild horses in Newfoundland.

Regardless of its exact range, the bison was undoubtedly an Appalachian native in the central hills, and one that immediately came under fire. Buffalo robes, with their exceptionally dense mat of woolly hair, were coveted for their superior insulation, and the meat was good eating, so the two-ton beasts were killed whenever the opportunity presented itself. Like the elk, bison frequented saline mineral licks, and hide hunters apparently knew about these traditional gathering sites. In 1806 one Thomas Ashe, an Irishman traveling in the United States, wrote of "an old man, one of the first settlers" in the highlands between the west branch of the Susquehanna and the Allegheny River, who built his log house near a buffalo lick.

"He informed me that for several seasons the buffaloes paid him their visits with the utmost regularity ... The first and second years, so unacquainted were these poor brutes with the use of this man's house, or with his nature, that in a few hours they rubbed the house completely down, taking delight in turning the logs off with their horns, while he had some difficulty to escape from being trampled under their feet, or crushed to death in his own ruins. At that point he supposed there could not have been less than two thousand in the neighborhood of their spring."

Ashe goes on to say that the settler and his companion killed between six and seven hundred bison the first two years, "merely for the sale of their skins, which to them were worth only two shillings each ...."

Ashe's tales of his time on the American frontier are sufficiently and baldly enough embroidered to make a reader cautious; at one point, he claims to have shot and wounded a black bear, which immediately treated its own injury by gathering clean leaves, "which with the utmost deliberation he stuffed in the wound and thus stopped the flow of blood." So when Ashe claims that multitudes of bison knocked down a settler's log cabin, or that the same settler killed between six and seven hundred buffalo—well, more than a grain of salt is required. A herd of two thousand bison in the Pennsylvania hills strains belief, as do stories of huge buffalo migrations near State College, collected from the families of the oldest surviving hunters by Colonel Henry W. Shoemaker at the start of the twentieth century—another source who rarely let strict accuracy stand in the way of a good story. There's no question that bison were present in the Appalachians, but outside of the western slope in Kentucky and West Virginia, they were uncommon at best.

The unremitting drain of hide hunting, coupled with a low population to start with, had rapid consequences, and by the Revolutionary War, bison were largely gone on the eastern slope of the Appalachians. The last in Pennsylvania was killed near Lewisburg in 1801 (or 1790, or 1810, depending on whom you believe), and the very last in the East, a cow and calf, were shot in 1825 near Valley Head, West Virginia.

For an animal said to have been so widespread, there is remarkably little left. No bison bones have been excavated from Indian middens in Pennsylvania, and although there were reports as late as the 1940s of buffalo coats, robes and horns dating from the 1770s, these artifacts are apparently gone, if they ever really existed.

If nothing else, a good Appalachian buffalo skin might clear up the mystery of *Bison bison pennsylvanicus*, a unique and possibly fictitious subspecies alleged to have lived along the Ohio River's headwaters. The Pennsylvania bison was supposed to have been bigger even than the Plains bison, "very

Captive bison seem to offer a glimpse of what once was, but bison likely were rare or absent everywhere except the western plateau regions of the central and southern Appalachians.

black, with short, crisp, curly hair, with some white about the nose and eyes, and stated to lack the hump ...." Our friend Colonel Shoemaker described the animal for science in 1915, more than a century after its presumed extinction, based not on specimens or eyewitness accounts, but on second- and third-hand descriptions—shaky science indeed. While anything is possible, mammalogists today are inclined to think that whatever bison lived in the Appalachians was the same subspecies found on the Plains, and for lack of any real evidence, the "Pennsylvania" form has been relegated to the taxonomic dustbin.

The nadir for Appalachian mammals came with the dawn of the twentieth century. The mountains were scalped and smoking, but among the slash piles and wildfire ashes, a new forest was sprouting. Deer, which had been so overshot that Pennsylvania had to import fifty from Michigan, found a paradise of thickets and responded with unbridled reproduction. Particularly in the last fifty years, the pace of healing has quickened as this new forest matures, producing some heartening changes in what were once thought of as "wilderness" animals.

For an example of how one species is adapting to humans—and the reverse—one should look no farther than the Poconos. Long before the undulating hills of this plateau became a honeymoon mecca, they were home to black bears, which by the middle of the century had been reduced to fewer than two thousand.

Then in the past several decades, development struck the Poconos like a whirlwind, and previously empty bogland and maple forest have been chopped up for vacation resorts, summer cabins and, most alarmingly, pricey second-home developments that sprawl across huge areas. Such events would normally bid fair to doom a large carnivore like the black bear. Oddly enough, that hasn't happened.

I got an object lesson in the adaptability of bears one afternoon when I spent several hours riding with Dr. Gary Alt. He cruised slowly in his green Blazer along the wooded lands of Hemlock Farms, a tentacled housing development in the forest about thirty miles west of Scranton. He was looking for trouble, in the form of a 250-pound female bear with three cubs and a penchant for raiding backyard barbecues. Alt, a slim, spectacled biologist who at the time was with the Pennsylvania Game Commission, seems an unlikely candidate for crawling into bear dens, but he probably knows more about Appalachian black bears than anyone in the world, having by that point studied them intensively for nearly twenty years. This particular bear was one of his research subjects, but she'd slipped out of her radio collar, and Alt wanted to find, dart and refit her.

The search, which went on all afternoon, failed to locate the errant bear, but riding with Alt, I began to understand what happens when you mix three thousand homes, ten thousand people, fifty-six miles of roads and twenty bears in just a few square miles. Far from being driven out of the woods by the development, the bears here have turned the change to their advantage. Trash can raids are common enough, but the Hemlock Farms bears have graduated to a more artful plane of pilfering; some have learned to zero in on the smell of charcoal, for example, knowing that burgers or steaks await them. They have grown accomplished at snatching bird feeders for the seed or suet, and groceries left sitting in an open car are fair game.

Some of the most brazen have even taken to breaking and entering. One woman lost a chest freezer's worth of meat (not to mention a garage door), only to be told by her insurance company that, since she could not prove the bear broke in with *intent* to steal, her policy wouldn't cover the loss. The company changed its mind only after the incident received national television coverage.

The bears are certainly pests, and Alt always points out in his lectures that each one is potentially dangerous, but Appalachian black bears are unbelievably tolerant of their human neighbors. Although there have been several fatal, unprovoked attacks in recent years, for the most part, the bears keep their awesome strength in check, even when they have reason to attack.

I had found that out for myself just a few hours earlier, while I was helping Alt check his trapline of foot snares not far from the development. Pulling up to the edge of a blueberry bog, we could hear the bawls of a young bear coming from the thick woods on the other side.

"Oh geez, we got a baby," Alt said as he climbed out of the vehicle with me, his assistant and two interns who'd come with me. "Sounds like a yearling. Its momma kicked it out this spring so it's alone, and now it's caught and scared and doesn't care who knows about it," he said.

Alt loaded a "jab stick," a long aluminum pole tipped with a hypodermic dart, with enough muscle relaxant to knock out a hundred-pound youngster. Then we trooped through the squishy bog, came to the edge of the woods and crawled under the overhanging rhododendron boughs.

We did not make a sound, and that was our mistake.

I was on my hands and knees, following Alt into the tangle, when I heard a powerful roar. Alt screamed a warning and threw up his arms. It was a gloomy day and the woods were dark, but when I looked past him I could see a huge, bristling black shape bounding toward us through the ferns—an adult bear, mouth open, just a few yards away and coming hard.

Alt and his assistant tumbled backward, yelling incoherently to scare the bear off. She skidded to so sudden a stop, so close to us, that the dirt she kicked up hit me in the face.

We abandoned the equipment and ran like hell through the swamp.

The error was ours in assuming that the yearling was still alone, as most second-year cubs are by June. With her offspring trapped and squalling, the sow bear had been wound to a fever pitch of worry, and our silent approach gave her zero warning. We were on top of her before she knew it, startling her into an attack.

When we went back a few minutes later, armed with a dart gun and making enough noise to raise the dead, she had melted into the forest, leaving us to drug, weigh and tag her 110-pound baby. The mother, as it turns out, was one Alt had been studying for fourteen years. This was her first aggressive action in all that time, and probably her last.

I was still a bit shaky several hours later, driving with Alt through Hemlock Farms. The homes in the development are expensive, many of them vacation getaways, although an increasing number house permanent residents willing to make the commute to New York City or Scranton. This, sadly, is the wave of the future in the Poconos, where the pace of development is forcing the native wildlife into closer and closer contact with humans.

So far, the bears have taken it in stride. While the mushrooming tract housing cannot ultimately be good for them, many bears have adapted to their new neighbors, perhaps sensing the opportunities that humans provide.

Such benefits go beyond food. By the road at many intersections in Hemlock Farms stand picturesquely jumbled piles of sandstone, moved years ago during road construction. Several were quickly appropriated by bears for winter dens. And not only rock heaps have been used.

"That house there," Alt says as we drive past an extravagance of cedar planking and wood decks, "had a female and two cubs under its porch last winter." Alt discovered the den by following the signal from the sow's radio collar, and found the happy family curled up under the joists of the deck, some of which the female had gnawed away to give herself more headroom.

"The owners didn't know they had a bear until we knocked on their door, and even then, they didn't believe us until we showed them," he said.

"They saw where she'd kicked rocks away from the foundation, but they thought it was raccoons." Such a den site is not unusual, nor is the complete ignorance of the human occupants. Black bears can be circumspect when they want to be.

Bears in the northern Appalachians usually head for their winter dens in November, occasionally earlier in the case of pregnant females, later for males; the Pennsylvania Game Commission uses this fact to time its bear season to spare fertile sows. Alt prefers the term "winter sleep" to hibernation, since the bears do not drop into the extraordinarily deep, deathlike state that characterizes woodchucks, bats or jumping mice. Instead, they drowse and doze with a somewhat reduced metabolic rate but can rouse quickly if disturbed.

During the course of the winter—early January in the Poconos—the females give birth to from one to five cubs. They are tiny, mewling creatures of less than a pound, covered in fine, black fur. Their eyes and ears are still sealed, and it appears they have little sense of smell in their first days, but they are acutely sensitive to warmth, and their mother's teats give off the most heat. Thus do blind babies and drowsy mother come together.

By the time the female leaves her den in late March or April, the cubs are lively, acrobatic toddlers of eight or ten pounds, capable of climbing trees swiftly. The female, lean from a winter's abstinence, is on the lookout for food—and so begins another season of conflict for the people and bears of Hemlock Farms.

As we drove the forested roads of the development, the residents, who know Alt simply as "the Bearman," flagged him down to tell him their latest bear story; the poor man has been hearing variations on this theme for years, but his interest seems genuine. It quickly became apparent that there are lots of bears in the neighborhood—one woman's garbage was hit that morning, a man's bird feeders were destroyed the night before, another said his wife looked up from the breakfast table a few hours earlier to see a yearling staring back at her. A road worker excitedly told us that we missed by minutes a big bear and three cubs, probably the family we'd been hunting.

Yet, for all the property damage caused by the bears, the residents seem remarkably acceptant of them. Bears are a fact of life here, and you get the feeling that for lots of these transplanted urbanites, having one in the shrubbery confers bragging rights.

Biologically, Pennsylvania bruins are unique; no other population of black bears in North America grows as fast, gets as large or has as many cubs per female. They are sexually precocious as well. In many parts of the West, a female black bear won't reach maturity until she's seven and a half or eight years old. In Pennsylvania, Alt has found, more than 80 percent breed before their third birthday.

Litters are outsized, too. Three cubs is the most common number, but four are as common as two, and five as common as one. In fact, more than 20 percent of all litters in the state consist of five cubs, a number that is unheard of in other parts of the bear's range.

"They're also growing very fast," Alt said. "A lot of our cubs weigh over a hundred pounds their first fall, and that's a figure that is rarely reached in other areas." Indeed; a western male in his third year averages about 75 pounds, while a male of the same age in Pennsylvania averages almost three and a half times that, around 260 pounds. Each fall hunters kill males between 650 and 700 pounds—the size of many central Alaskan grizzlies. In 1992, a man hunting in the Poconos shot a male black bear with an estimated weight of 827 pounds, the most massive black bear ever recorded.

It all comes down to food. The Appalachians hereabouts are a table set for the eating, with mountains of mast-producing oaks and beeches and wetlands full of wild berries, skunk cabbage, tussock-sedge and other delectables.

Thanks to the Wisconsin glaciers that scraped across this region, bogs, swamps and wet meadows are abundant in the Poconos. One of the most startling results of Alt's work has been the realization that black bears here are predominantly wetland animals. The wooded swamps and bogs of the Poconos, tangled with blueberry and rhododendron, make up less than 2 percent of the land area, but are where the bears spend more than 70 percent of their time. And because they are all but impenetrable, the swamps provide absolutely essential escape cover.

Hemlock Farms was built in and around a number of wetlands, including Maple Swamp, while lakes, ponds and bogs dot the surrounding countryside. The area is a maze of soggy land, so that twenty yards beyond a neatly manicured lawn may be a sphagnum sink where people rarely venture.

Alt parked the car and walked between two beautiful homes to the edge of the woods. The ground immediately became sodden, and my boots sunk to the ankles, then the shins; ostrich ferns rose to shoulder height around us as we threaded along a narrow trail that I would not even have noticed were it not for Gary's guidance.

The hemlocks and rhododendrons closed around us like a wall, even though we were within a long stone's throw of several homes. The ground became higher and drier, and even I could see the path now, worn in the red-green sphagnum. Every eighteen inches or so, there was a slight depression in the ground the size of a bear's foot; Alt explained that bears habitually step in exactly the same place each time they walk a trail. Another few yards and the path ended in a tiny clearing just big enough for both of us, filled with what looked, at first, like a giant, flattened bird's nest.

In a moment I realized it was a bear's day bed, about five feet across, carefully constructed of tiny twigs bitten off and carried here to make a raised, dry sleeping place. There was a musty odor, and I got a chill down my spine and an unmistakable feeling of trespass. I knew from long experience that these are tolerant, passive animals, but the morning's scare was still too fresh in my mind.

Pennsylvania's black bears are doing amazingly well. Despite a human population of more than 12 million, the state now supports more than fifteen thousand black bears, despite increasingly liberal hunting seasons and an annual kill that often exceeds four thousand animals. The reasons for this population explosion include better protection when numbers were low, abundant food and the bears' own amazing fecundity, but some credit must also go to the growing adaptability of bears to humans—and vice versa.

In recent decades, black bears have been moving out of their old strongholds in the mountains and into areas where they haven't been seen (in some cases) for centuries, like the suburbs of Philadelphia. They are now common in and around urban hubs like Scranton, Wilkes-Barre and State College—as well as northern New Jersey, and the suburbs of New York City.

For a while, bear numbers in the southern Appalachians lagged well behind the recovery seen in the north—small populations isolated in fragmented habitat, competing with invasive wild pigs for food, and under pressure from poachers that shot them for their gall bladders, used in traditional Asian medicine. But with a crackdown on poaching, there has been a spectacular turnaround in the southern mountains, too, with soaring bear populations—and, increasingly, the same conflicts between people and bears that the rest of the Appalachians have already experienced.

It may seem strange that, while bears are doing well in much of the Appalachians, a rat is not. We tend to think of rats as the ultimate survivors, able to tolerate any circumstance, and that's true of the Norway rat, the all-too-common denizen of city alleys and country barns.

But the Norway rat is not native to North America; along with the black rat and the house mouse, it tagged along with the colonists streaming out of Europe. The Allegheny woodrat, on the other hand, is a native and a long way from the usual conception of "rat." It is a round-faced, large-eyed creature with fawn coloration (warm brown above, white underneath and on the feet) and a furred tail with the same bicolor pattern—an altogether engaging animal more like an overgrown deer mouse than a rat. The species that inhabits the Appalachians lives in rockslides and boulder fields from southern New York south; another species, the Florida woodrat, lives in lowland habitats across the South.

I got to know woodrats when I was a teenager, scrambling over the rocky ridgelines near my home and exploring the shallow caves formed by the boulder slides. I often found their nests, large mounds of sticks, leaves and shredded bark, tucked in protected recesses and surrounded by remarkable assortments of junk—everything from bits of rusty metal and old shotgun shells to the jawbones of deer and strips of plastic surveyor's tape. Woodrats, like their more famous relatives out West, are "pack rats," possessed of an inexplicable urge to collect anything out of the ordinary, especially anything bright or shiny. Legend has it that a pack rat always leaves something behind in trade when it pilfers an item. Biologists say what usually happens is that the rat, its mouth already full, drops what it was carrying to take away the new bauble.

I rarely saw the rats themselves, because they are an almost strictly nocturnal species, but on a few occasions, I would bump the nest with a stick and be rewarded with a wide-eyed face staring back at me from the entrance hole, framed by two round, pink ears and a halo of twitching whiskers.

Woodrats are generally solitary at home, keeping to their own dens except in breeding season or when rearing young, but they live in loose colonies, and even share latrine sites, for these are fastidious animals. I once watched one perch at the edge of a drop-off and laboriously groom itself, licking and smoothing its fur with the intensity of any cat, a performance that continued until the evening grew too dark to see any more.

Woodrats suffer guilt by association, so much so that their supporters have tried verbal gymnastics in their defense. A friend of mine, a wildlife biologist for the state, makes it a point to refer them as "Allegheny cave squirrels" whenever possible to distance them from Norway rats (they are members of an entirely different family, the cricetids, to which all native mice and voles belong). Others have gone so far as to dub them "long-tailed boulder bunnies" or "Pennsylvania pikas." Supporters also like to mention that the Indians had none of our qualms about the woodrat; it is said to taste better than squirrel, and its bones are a common component of their trash middens.

Tastes change, and I know of no one who's been tempted to sample one themselves. Living up in the mountains well away from human development, with no food, commercial or sporting value, the woodrat would seem to have no worries. And yet, inexplicably, the cave squirrel began to vanish, slowly and then with speed and completion. By 1980 they were largely absent from Pennsylvania east of the Susquehanna, a region where they had been abundant; they also became extinct (or nearly so) in New York and Connecticut, and virtually extinct in northern New Jersey and Ohio. Only in the southern portions of their range, from West Virginia to Tennessee, have populations remained relatively stable.

For years, the disappearance of the Allegheny woodrat has been one of the most baffling puzzles for Appalachian biologists. The decline was so precipitous, and it occurred over such a wide area, that none of the usual causes—habitat destruction, for instance—seemed to fit. The few scientists who bothered studying woodrats assumed the cause had to be related to food supply, and the timing of the crash came within a decade of the first appearance of gypsy moths in the region. Biologists knew that the stress of gypsy moth defoliation reduced crops of acorns, an important food for woodrats.

Perhaps, it was theorized, the dwindling food supply hurt woodrat reproduction and forced the survivors to spend more time foraging for alternate foods away from the security of their rock piles. This may have led to increased predation by owls, foxes and weasels, and a further erosion of their numbers.

Others argued that the colonial nature of woodrats, which live in disjunct pockets of habitat rather than a continuous range, might be the culprit. The habitat between colonies has changed with time, and highways, fields and utility corridors are hostile ground for a young woodrat striking out for new territory. Others felt disease might be a factor; still others that logging was to blame, even though many of the extinct colonies were miles from timbered ground.

Competition from other animals was considered, and here the major suspect was the raccoon, which had become exceedingly common in the central Appalachians in the previous twenty-five or thirty years. Raccoons are omnivores, but they eat many of the same foods favored by the vegetarian woodrats, including berries and wild grapes; they are also large and aggressive and prey on smaller mammals, including woodrats. Many naturalists agreed that raccoon populations on the high ridges increased after the gypsy moths arrived, probably because the forest canopy was opened up in many areas, allowing more undergrowth. Could the raccoons simply have elbowed the woodrats out of their niche and taken it for themselves?

The debate raged in the small circle of woodrat researchers for more than a decade, with no real resolution. Ironically, when a key answer finally appeared, it came from a completely unexpected culprit—*Baylascaris procyonis*, a parasitic roundworm.

Baylascaris is a nasty creature, infesting the digestive tracts of raccoons, which pass its eggs along with their feces. Because raccoons eat a lot of fruit and pass the seeds whole, their dung is often recycled by seed-eating rodents. The parasite does not seem to unduly trouble the raccoons, but when the eggs are ingested by any other mammal or bird, the larvae that hatch from them wander aimlessly through the host's body, eventually ending up in the brain. The result is death, and no treatment is known. There have even been human fatalities, generally young children who picked up roundworm eggs on their hands and then stuck them in their mouths.

Woodrats, with their penchant for collecting things, routinely pick up dried animal scat to add to their middens. By carrying raccoon feces in their mouths, they would almost certainly become infected with roundworm eggs.

Suddenly it all made sense. A deadly, fast-acting agent had been identified, and the timing coincided with the arrival of large numbers of raccoons in woodrat habitat. The roundworms may have been the final ill in a series of blows. Gypsy moth damage made winter food supplies of acorns less stable, while habitat fragmentation cut off colonies from one another and increased the number of predators like red foxes and house cats that avoid deep woods. And finally, raccoons infected many colonies, especially the lower, more fertile sites most attractive to young, wandering woodrats — the very individuals needed to maintain colonies.

The bird banding station I run each fall is on a boulder field that once held a colony of woodrats, a colony that died out more than fifteen years ago. Last year, however, on a slow day when I had some time to explore, I came across a pile of small, oblong scat on a flat rock. The droppings were too big for a chipmunk's and the wrong shape for a gray squirrel's, and there were so many of them that the place was obviously being used as a latrine. I tried not to get my hopes up, but I kept looking. When I brought up a friend who's an expert on woodrats, he found some suggestive sign, too, but nothing conclusive.

This autumn I spent several hours combing the rocks, sticking my head in crevices and shining a flashlight between boulders. Finally I found what I'd been hoping for—a nest of sticks and twigs and leaves about the size of a bushel basket, with the fresh shells of acorns scattered around. The cave squirrels had come home.

If the nineteenth century was the bleakest time for the Appalachians' large mammals, the twentieth was somewhat kinder, thanks to more enlightened game management and maturation of the forests. Like the black bears in the northern ranges, other animals previously restricted to wilderness have made the most of the improving situation.

In New England, moose are a common sight again—not just in Maine, but in Vermont and New Hampshire, where they were completely extirpated. Moose have reestablished themselves in smaller numbers in the Adirondacks, and a few have wandered across New York (one to within a few miles of Pennsylvania), others into the Berkshires, Connecticut, Rhode Island and even coastal Massachusetts.

Much as bears, these are monstrously big creatures that can, like will-o'-the-wisps, live near humans without ever being seen. A good friend of mine in western Vermont finds their softball-sized tracks each winter in the snow around his lakeshore home but has yet to see the moose themselves. It frustrates him to think that a thousand-pound mammal can blend so completely into the woodwork.

Even though moose are doing so well in Vermont that the state instituted its first moose hunting season in a century in 1993, the Green Mountains, with their sugar maple stands, are really better suited to deer. Moose don't come into their own until you move north, especially to central Maine. Here, Mount Katahdin lords over the lesser peaks of the Longfellow Mountains like a bunched muscle, rearing up from the flattish coastal plain of spruce and birch that stretches two hundred miles to the sea. This is a boreal jungle of regrown clear-cuts, thick spruce blowdowns, open heaths and impenetrable bogs, a semi-wilderness seamed with anonymous dirt logging roads in which it is possible to become quickly and seriously lost.

Smack in the middle of Maine's North Woods is Baxter State Park, more than three hundred square miles of forest with the gray bulk of Katahdin at its core, donated to the state in the 1930s by former Governor Percival Baxter, with the provision that the park remain forever wild. There is no hunting allowed, and the moose herd has done well over the years—probably a little too well, since there are no wolves to keep them in line. It is almost impossible to spend much time in Baxter without seeing at least a few of the great beasts.

Sandy Stream Pond, a jewel of a lake set under the walls of South Turner Mountain, is especially popular with moose, since its muddy bottom supports a lush growth of aquatic plants that supply the moose with sodium, a mineral lacking in their normal diet of twigs and woody browse. I had walked the easy trail back to the lake and was sitting on the shore, hunched over my tripod, watching through the camera viewfinder as two bulls and a cow methodically sucked down water lilies a hundred yards away.

One male, the largest, would submerge almost completely, with just the peak of its hump breaking through the choppy waves; one could easily believe it was just a dark rock. But then, after thirty or forty seconds, the massive head would break the surface like a breaching whale, cascades of water pouring off the wide, palmate antlers in a flickering halo. It was like watching a sea serpent, or the incarnation of the old Ojibway legend of the evil Manitou with the rack of a moose and the body of a bear that rose from the depths of lakes to kidnap women.

I was transfixed, nothing moving for the longest time but my hair ruffled by the strong wind. Then, like the tickle of a mosquito buzzing just beyond consciousness, I got that creepy, hackle-pricking feeling that I was being watched. Without rising, I turned my head slowly to look behind me, and found myself peering into two enormous nostrils just a foot or so away from my face.

No doubt puzzled by this odd rock where none had been before, a yearling bull had inched closer, the soft sounds of his approach masked by the gusting wind. I stared up at his looming bulk and jerked spasmodically, and he in reaction blew a startled snort through that flabby nose and backed up a pace. I stayed motionless; he ambled quietly way. I think we were both a little embarrassed by the encounter.

It is obligatory, I suppose, to say something mocking about a moose's appearance, but it is mostly adolescents like this young bull that look like an unfinished thought. The young calves have that wide-eyed innocence of babyhood that forgives any misproportion, but the yearlings are a mess: their too-long legs, their heads too big for their bodies, the ridiculous knobs of the bull's first tiny antlers and the pathetically thin bell of flesh and hair dangling like a rope of aimless drool under the chin. They look like a self-parody, as though someone started with a rather ordinary-looking animal and tweaked the features too far, like a cartoonist lampooning a politician. Nixon's nose was never *that* big, and even from the range of a foot, I can hardly believe a moose's is, either.

That gawkiness disappears with adulthood. A bull in the breeding season is one of the most magnificent big game animals in the world; the winter coat is chestnut on the hump and neck, deepening to a light-swallowing black on the belly and flanks, except for the faintest hint of iridescent purple when the sun catches it right. Maturity and the rut fill out the neck and the chest with muscle; the bell thickens to a heavy, patriarchal beard; and the cleaned rack gleams at its polished tips. One moment the moose stands like something carved of black onyx and shadows, then gathers itself to move, the peculiar, high-stepping trot like a Tennessee walking horse's, unimaginably fluid for so massive an animal.

As moose have rebounded in the northern Appalachians problems have unavoidably followed. The most serious are car collisions; even a solidly built sedan stands little chance when it slams into twelve hundred pounds of moose, and a little compact crumples like paper. New Hampshire has taken to erecting strong warnings along with the more usual Moose Crossing signs. On Highway 3 north of Pittsburg, a veritable mooseway along the upper Connecticut River, signs state the problem baldly: "Brake for Moose— It Could Save Your Life. 195 Collisions." The efficacy of such warnings is unclear; as is so often the case with large, somewhat tame mammals, people tend to treat moose like domestic livestock or safari-park inmates, gawking at the first few they see, then barreling past others as the novelty wears off.

The moose have come back by themselves, but not every animal can—or we lack the patience to allow them to do so. There's a growing trend toward restoring some of the species that have been lost to time. Its roots lie with the men who trucked Wyoming elk to Pennsylvania in 1913, and while the methods have grown more sophisticated, the fundamental job is the same: take spare animals from a place where they're common and release them in a place where they aren't.

Predictably, results vary. New York successfully moved fishers from the Adirondacks to the Catskills and tried to bring lynx back to the High Peaks, although some of the radio-collared cats wandered as far as Pennsylvania and New Jersey, and the program fizzled a few years later. But fisher reintroduction in Pennsylvania was so successful that, twenty years later, the fox-sized weasels are found across most of the state. Peregrine falcons were "hacked" back into the wild by rearing chicks on isolated cliffs in New York and New England, where they now breed naturally—though in much of the Appalachians, the falcons have become largely urban raptors, nesting on bridges and skyscrapers.

In the 1960s woodland caribou were released in Baxter Park, where the species had been eliminated in the early 1900s. The experiment flopped when the migratory animals simply wandered away and died, but the dream of bringing caribou back to the state stayed alive. In the mid-1980s a privately funded effort based at the University of Maine in Orono embarked on the same goal, using a technique pioneered in Quebec. Nonmigratory caribou were to be captured on the island of Newfoundland, held in outdoor pens at the university as breeding stock and their progeny released in Baxter.

Phase one, the captive breeding, appeared to go well, and by 1989 the first releases were made. That year and the next, more than thirty caribou were moved to Baxter, where disaster struck. Instead of staying on the release site in a herd, the caribou scattered like quail, falling victim to bears and coyotes. Even worse, they had, unknown to the scientists, become infected with brainworm back at Orono, where no one had thought to wonder whether the deer that had previously lived near the caribou enclosures might have left behind a lethally high density of the parasites. Those caribou that managed to evade the predators fell victim to the worms; one of the last, a female, toppled off the sheer cliffs of Pinnacle Ridge, her senses warped by the parasitic load. For the second time in thirty years reintroductions were shelved. Another attempt, this time bringing wild caribou directly from Newfoundland, failed because bears and coyotes again made short work of the transplanted animals.

Caribou are one thing, but predators are something else again. There is no concern that a large ungulate is going to graze your children to death, but the Red Riding Hood mythos is lodged deeply in our culture, and biologists have had a hard time convincing the public that it's safe to restore lost meat-eaters, even in huge wilderness systems—witness the long-running tempest over the reintroduction of wolves in Yellowstone.

So it is strange that no similar furor accompanied the first tentative release of wolves in the Appalachians. Perhaps it is because the red wolf doesn't look like something that would eat little girls. The smallest of North America's two native wolves, this lean, rangy canine weighs no more than sixty or seventy pounds, with a leggy look and a close-cropped coat that can be reddish but is more often grizzled like a coyote's pelt. A hunter more of rabbits and muskrats than big game, it was once found from the Mason–Dixon line to the Florida Keys and central Texas.

Although not a habitual stock-killer, *Canis rufus* was tarred with the same brush as the gray wolf, and as assiduously hunted. It vanished from the Appalachians by the turn of the century, and by World War II was restricted to pockets of thick bottomland of the Gulf Coast and the lower Mississippi basin.

It was an untenable position, with small, increasingly inbred populations cut off from each other—and the situation got dramatically worse by the 1950s, when coyotes began to usurp the wolf's old range. Coyotes and red wolves are close in size and appearance and extremely close in genetic makeup, so hybrids between the two species are fertile. With few of their own kind left, the remaining wolves mated with coyotes, and by 1970 the tiny enclaves of *C. rufus* had been swamped. Only a last-ditch effort saved the species by live-trapping everything in Texas and Louisiana that resembled a red wolf, then carefully winnowing the hybrids from the real McCoys, based on physical characteristics. By 1975 the biologists had identified seventeen "pure" red wolves, which along with those already in captivity, would form the nucleus of a breeding program.

Bigger and more powerful than western coyotes, those in the East have proven to be another beast altogether—an amalgam of coyote and timber wolf, and so adaptable that they are now found in every county east of the Mississippi.

Hauled across the continent to Washington State, where they became wards of the Point Defiance Zoo, the forty-two original red wolves proved to be fairly prolific breeders, eventually more than tripling their numbers. By the mid-1980s biologists were ready to try experimental releases, choosing isolated sea islands off the Carolinas, Florida and Mississippi. At Bulls Island in South Carolina, wolves mated and raised pups, and all but one even managed to survive the ferocity of Hurricane Hugo in 1989.

Biologists knew that the offshore locations were too small for sustainable wolf populations, but they made ideal breeding sites, allowing captive-bred wolves to produce wild pups for eventual release elsewhere. In 1986 the project moved to Alligator River National Wildlife Refuge in North Carolina, on the mainland, for the first true restoration to the wild. After spending nearly a year of acclimatization in large outdoor enclosures, two wolves were released on September 14, 1987, and thirteen more eventually followed—the first time an American animal extinct in the wild had been returned to part of its range. The wolves survived and litters of young followed, but every bit as crucial as breeding success was public support. Alligator River became something of a tourist attraction, and the locals, knowing a good thing economically as well as ecologically, later picked the red wolf as their town symbol.

It is one thing to release wolves in a little-known coastal refuge or on inaccessible sea islands. It is quite another to do so in Great Smoky Mountains National Park, with its 10 million visitors a year and its reputation as America's busiest piece of public real estate. Yet in January 1991 paired wolves were moved to the park's Cades Cove region for their long settling-in period, and in November of that year, the doors opened. For the first time in far too many years, there were wolves in the Appalachians.

Or were there? At the same time that public interest was focused on the Smoky Mountains project, a team of scientists announced that DNA analysis of old pelts and blood samples suggested the red wolf was nothing more than a hybrid between the gray wolf and the coyote, with no unique genetic material of its own. This raised a number of disquieting questions: If the red wolf is not a valid species, should it qualify for Endangered Species Act protection and funding? Should it be restored to its former home at all, especially since coyotes have since moved in?

The U.S. Fish and Wildlife Service quickly said that, whatever the animal's origins, the red wolf project would not be abandoned. A number of biologists also pointed out that even if the red wolf is a hybrid, the conditions that led to its genesis must have been unique, for no others like it have arisen in areas where gray wolves and coyotes overlap.

The issue, more than twenty years on, remains unsettled. Dueling scientific studies have suggested, on one hand, that red wolves and the smaller eastern wolves of southeastern Canada (known variously as timber wolves or Algonquin wolves) are essentially widely separated populations of what was once a single, continuous species. But more recently, other researchers have looked at the genetics of dogs, wolves and coyotes and concluded that, as once suspected, red wolves are largely coyote in origin, with the remaining 20–25 percent of their genome contributed by gray wolves.

In any event, this much is clear. In the untouched wilderness of the southern Appalachians, in the millennia before the forests fell, the predator at the peak of the food chain was what we call the red wolf. However it got there, whatever its ancestry, it filled a specific historical and ecological niche that no other near-relative could quite fill. But the world is not what it once was, and after all the fanfare of its Smoky Mountains return, the red wolf couldn't survive in this, one of the largest tracts of wild land in the East. Of the thirty-seven wolves released in the half-million-acre park, and the thirty-four pups born there, almost all died from starvation, disease or when they wandered from the park and were hit by cars, shot or poisoned. Nine years after the effort began, the last four survivors were retrapped.

Wild red wolves still hunt the lowlands of eastern North Carolina, but for wolves in the Appalachians, the focus has shifted north. A few wolves have been documented in Maine, although the consensus is that these were released animals, not eastern wolves from the wild population in Quebec. Other, sketchier reports of wolves in Vermont, New York and New Hampshire may also be released captives, or wandering Canadian wolves. There is deep ambivalence about restoring wolves to the northern Appalachians, and if they come back, it will have to be on their own four feet, like the handful of pioneering wolves that have reached New Brunswick and even Newfoundland.

But four feet can carry a wild animal a long way. In June 2011, a 140-pound mountain lion was killed by a car in southern Connecticut. Subsequent genetic tests, coupled with eyewitness sightings, photos and scat samples stretching across the East, showed—to pretty much everyone's shock – that the cat had walked from the Black Hills of South Dakota, fifteen hundred miles to the west.

Nature has a way of filling vacuums. We may never have herds of bison stamping their wide paths beneath the autumn-flamed hardwoods, their hoofbeats drumming a distant, dimly heard thunder, but the silent cats and the howling wolves may yet return to join the bugling elk, giving voice to mountains struck dumb for silent centuries.

On the tree-lined main street of Lancaster, New Hampshire, a small town in the state's northern panhandle, there is a statue of a wolflike animal perched atop a boulder in the village green, between the library and a church. Below it is mounted a brass plaque:

1763 – 1913
To honor the brave men and women who redeemed Lancaster
from the wilderness, this memorial was erected by their
loyal sons and daughters on the 150th anniversary of the
founding of the town, July 6, 1913.

One could certainly argue with the language, but "redeemed" is so hugely arrogant a word I almost find myself forgiving the past notables of Lancaster their hubris; those were heady days before the Great War, with a continent newly tamed and nothing in sight but a burnished future. I'm more intrigued by the wolf itself.

Given the context, I would have expected something in the old Victorian, Nature-red-in-tooth-and-claw mold, a snarling wolf epitomizing the wilderness that the early Lancastrians had so bravely redeemed. Instead, the sculptor caught the animal in midstride with a forefoot lifted, its head down and canted right, its tail hanging loosely. The muzzle may be a bit too pinched and doglike, but the overall effect is strikingly realistic and, given the context, disturbingly somber and sympathetic. All unintended, this is a memorial to the vanquished, not a paean to the victors.

I can't help but wonder if the folks in Lancaster, back in 1913, weren't just a little disappointed with it.

Flint Projectile Points

Chapter 7

ROOTS IN THE HILLS

Many years ago I interviewed a local man who had made a reputation as an amateur archaeologist. He had no formal training—indeed, he worked on a state highway department road crew—but he knew his subject and had a knack for finding Indian artifacts, and his opinions carried weight with academics.

We'd been driving rather aimlessly through the countryside while we talked, among the folded hills of central Pennsylvania, when he mentioned that finding an ancient campsite is child's play. Needled by his nonchalance—I'd been looking for arrowheads for years without success—I challenged him to find one.

He sat staring out my car window for quite some time, then suddenly told me to pull over. We stopped in a thin band of woods, with cornfields beyond to the south and a low hill to the north. A spring rose from the side of the hill and joined a small stream below. He hopped out of the car without a word, strode up the hillside to a small flat, and scuffed with his boot in the soil.

Less than two minutes later he uncovered a broken projectile point, the notched butt section chipped from flint and perhaps an inch and a quarter long. He handed it to me without a word, and I brushed the loam from it, feeling the slick smoothness of the flint and seeing the smoky translucence of the rock.

It was, he later explained, merely a matter of looking for the right conditions. The hill to the north would have given shelter from the wind, the southern exposure sunlight, and the spring was a ready source of clean water. All three elements, combined with a flat spot for bark shelters, made it an ideal camp for hunters or travelers.

"You have to remember that there were people around here for thousands of years," he said simply. "It's almost inevitable that someone stayed here—not just once, but many, many times."

The Appalachians have been home to humans for longer than most of us realize; both the Native and the European periods stretch back farther in time than the average person appreciates. And more than perhaps any other mountain system in North America, the Appalachians have been changed by—and changed—the human cultures that sheltered among them.

Still, in the grand scheme of geologic time, our species is a newcomer to these hills, present only for a few thousand years out of a span of hundreds of millions. No one can pinpoint the first visitors, of course, but until recently most experts agreed on a general date: sometime after 11,500 years B.P. (before present), the time mankind was thought to have first crossed from Asia.

At that time, with the world gripped by continental ice sheets and global sea levels subsequently lowered, humans would have been able to cross the exposed Beringia land bridge between Siberia and Alaska. From there they would have moved south, following a narrow, ice-free corridor between the Cordilleran ice sheet along the Pacific coast and the Laurentide ice sheet to the east—a thin waist of land running through what is now British Columbia and Alberta. From there these Paleoindians spread out, colonizing the rest of North, Central and South America. These first immigrants are known as the Clovis culture, named for their characteristic, elegantly fluted projectile points, first found near Clovis, New Mexico.

The archaeological evidence certainly backed up that theory. None of the thousands of excavations in the western hemisphere, from which bone or charcoal could be tested radioactively for age, violated that earliest date of about 11,500 B.P. It was a neat, simple answer to the old questions of where and when humans arrived on this continent.

But that neat, simple answer has been swept away, thanks to findings at sites as widely scattered as the Yukon, Virginia and Chile. The Meadowcroft Rock Shelter south of Pittsburgh, in the rolling Appalachian Plateau country of western Pennsylvania, is one such intriguing location. The shelter, a deep hollow beneath a sandstone overhang, has been in the same family since 1795. In 1969 the owner, Albert Miller, uncovered stone chips, bone fragments and mollusk shells—enough to convince him that Native Americans had inhabited the cave, and enough to lure archaeologists from the University of Pittsburgh into starting a five-year dig in the shelter.

Their findings were eccentric, at least by the accepted wisdom of North American archaeology. A piece of bark, once probably part of a basket, was radiocarbon dated at between 16,200 and 19,000 B.P. Furthermore, the projectile points at the lowest levels matched—at least according to Meadowcroft's supporters—the style of blades used in Asia roughly twenty thousand years ago.

Detractors claim the Meadowcroft samples were contaminated by older coal dust washed in from other areas, something supporters have rebutted based on an absence of a coal source and the level of the water table. More tellingly, critics also point to fossil pollen, which suggests an oak forest at the time the artifacts were buried—an unlikely habitat for a region that was then less than one hundred miles from the nearest glacier.

If Meadowcroft was the only anomalous date, it would probably have been ignored as a fluke, but it was only the vanguard of many sites that finally brought the "Clovis first" orthodoxy crumbling down. No others are in the Appalachians, but from the Bluefish Caves and Old Crow sites in the Yukon, to Monte Verde in Chile, Cactus Hill on the coastal plain of Virginia, the Channel Islands of California and Page-Ladson in Florida, archaeologists have uncovered ever-more compelling evidence of pre-Clovis cultures in the Americas.

Genetics and linguistic analyses have added to the picture, suggesting multiple waves of human immigration into the New World, perhaps as early as thirty thousand years ago. The earliest pioneers may well have walked across the Bering Land bridge, as once was assumed, but they could also have been seafarers, following the fertile tidewaters around the north Pacific coast and all the way to Tierra del Fuego, and landward to the east.

About five hundred miles south of Meadowcroft is another rock shelter, Russell Cave in northeastern Alabama. It, too, lies among the low hills of the Appalachian Plateau, in one of the fluted cove valleys that lead to the Tennessee River. While not as old as the projected dates for Meadowcroft, Russell Cave boasts one of the longest histories of human occupancy in North America, an impressive nine thousand years of inhabitation.

I came to the cave on a windy spring day to find two yellow school buses in the parking lot of the National Monument visitors center. Hearing the hubbub of kids down by the cave, I took a long walk up into the empty hills instead, through a stately forest of oak, hickory and locust. Tufts of blue phlox grew along the trail, with the white spotlights of mayapple flowers shining down on them from beneath their green umbrellas. Here and there, limestone outcroppings protruded from the leaves, many nodules of chert embedded in them—a handy source of raw material for cultures dependent on stone tools.

Russell Cave has been probed and dug for years, starting with amateur archaeologists in the 1950s and culminating with digs sponsored by the National Geographic Society and the National Park Service in the late fifties and early sixties. The findings fit neatly into the old 11,500 B.P. format for human arrival in the New World. They show succession of Native cultures, starting about nine thousand years ago when the rock face partially collapsed, forming the current cavern.

The earliest culture, represented by a single spear point, is a phase known as the Transitional Period, so named because it linked the Paleoindian cultures of the ice age with more modern, postglacial societies. Many more remnants were found from the Archaic culture that followed, from 9,000 B.P. to 2,500 B.P. They appeared to use the cave as a winter shelter for an extended family or small band that camped here to hunt and escape the worst of the weather. Their dead—some of them, at least—were buried in shallow graves scooped out of the cave floor. The bones of their prey species were common in the buried layers of the cave—mostly deer and gray squirrels, turkeys, bears and raccoons. The age of the many deer found in the cave, based on bone length and tooth pattern, adds weight to the idea that it was a winter camp. Most were about a year old when they died, which is what one would expect if they were killed in late winter.

The abundance of squirrel bones, along with other evidence, strongly suggests that the forest around Russell Cave nine thousand years ago was similar to what is there now, a hardwood community heavily laced with oaks. But there were surprising differences. Porcupines, now found no farther south than West Virginia, inhabited Alabama then, as did a species of now-extinct peccary. In some respects, the effects of the ice age still lingered.

Far down the slope, I heard the rumble of the school buses churning to life, so I turned down the trail to the cavern. The caves—there are two openings side-by-side, one lower and flooded with spring water, the other higher and dry—are surprisingly big. They formed as a result of a limestone sink, when an area the size of a small parking lot subsided, exposing a spring that flows a short distance and enters the flooded cavern.

The water flowing from the spring was somewhat milky, and barn swallows were dipping into it for drinks, making little sunbursts of reflection on the water, then disappearing into the black hole of the cave to their nests. A tiger swallowtail butterfly drifted aimlessly in and out of the opening, flashing gold in the sun, winking out for long moments in the shadow, then reappearing, weaving among the thin tendrils of creeper that hung down over the face of the cave.

Inside the air was cool and damp, and my eyes took a minute to adjust to the gloom. The living floor of the cave has been dug away by years of archaeological excavation, and you can climb down wooden steps beside the neatly cut earth to see the horizon lines—the distinct layers, lettered A through G from latest to oldest, that represent different cultures. Here, from their trash and their bones and castoffs, scientists trace the human story of Russell Cave.

The litter of the Archaic tribes is found in the lowest layers, E, F and G; above them, their culture is completely replaced by another, the Woodland, starting about twenty-five hundred years ago. Rather than using the largish spear points favored by Archaic hunters, the Woodland people chipped smaller, delicate points, presumably for hunting with bow and arrow, thought to have become the weapon of choice by this time. Pottery appeared, and here (as at other Woodland sites in the Appalachians) storage pits for acorns came into wide use. There are neatly worked bone tools and ornaments of shell and bone—all indications of a society in which there was more time for the finer things in life, instead of the constant, minute-by-minute struggle for survival.

This period of Native history, which lasted until about A.D. 1000, saw the blossoming of agriculture and the decline in the use of Russell Cave as anything but a temporary winter hunting camp. The focus of life had become the village, with its cultivated fields and steady, dependable food supply. Still later cultures—the Mississippian people and tribes of the historic period like the Cherokee—used the ancient rock shelter hardly at all.

I walked back into the sunlight and startled something small and furry near my feet. I waited patiently, until finally a flat, brown woodchuck head rose above the waterleaf and ferns. Then another, and another, until five youngsters the size of big kittens stood staring at me. Their den, I noticed, was dug beneath a sign warning, "Embankment and bluff area closed due to safety and erosion."

By the time Europeans arrived in North America, Native cultures in the Appalachians had evolved a sophisticated mix of agriculture and hunting, with an astonishing diversity of societies up and down the mountain chain. In fact, the subject is immensely complex, made all the more so by the constant shifts in tribal territories, migrations and warfare.

There were hundreds of tribes, divisions and bands, jockeying constantly with each other in outright warfare or more subtle incursions of migration and trade. To the north were the Beothuk in Newfoundland (the Inuit had retreated farther north into the Arctic) and the Micmac, Passamaquoddy and Abenaki in the Maritimes and northern New England. The Iroquois confederacy, which united the five tribes of the Seneca, Cayuga, Oneida, Onondaga and Mohawk (and later a sixth, the Tuscarora) controlled the New York region.

In the ridges and valleys of the central Appalachians we know about the Lenape (Delaware) and the Susquehannock of Pennsylvania, but virtually nothing about the "Monongahela people" and other poorly understood cultures of the Ohio drainage from western Pennsylvania south through West Virginia; they seem to have vanished before whites arrived, perhaps (archaeologists suspect) from war with the Susquehannock or the Seneca, perhaps from cultural assimilation, perhaps from the devastating epidemics that spread rapidly and fatally across the East after contact with Europeans. And still farther south, centered in the Smokies, were the Cherokee, one of the largest and, as it proved, most cohesive tribes in the Appalachians.

It should be noted here that although many tribal territories encompassed the mountains, the Native Americans—like the whites that followed them—generally preferred to live in the valleys; the higher and more rugged sections of the mountains were largely uninhabited or were used only for periodic hunting excursions. The reason was simple expedience—why trudge through steep, rocky peaks and ridges when the level lowlands provided everything a village could want? Besides, there were dangers for travelers of any race in the mountains. In many areas, summer travel was often done by canoe, to avoid the abundance of timber rattlesnakes that haunted the rocky slopes. While the Cherokee homeland had always included the Smokies, for instance, it was only with the steady incursion of Europeans that they found themselves pushed back into the high mountains themselves.

In most minds, European settlement of the New World dates from the sixteenth century, but in truth, the date is much earlier—earlier even than the Basque whalers who made Labrador and northern Newfoundland their base in the late fifteenth and early sixteenth centuries. In the year A.D. 985 or thereabouts, a Norseman named Bjarni Herjolfsson (or Herjulfsson) left Norway to visit his father, who had emigrated to the Viking settlement on Iceland. When Bjarni arrived at Iceland, however, he discovered that his father had continued on to Greenland, which had recently been colonized by Eirik the Red. Foolishly—for no one in his crew had even been to Greenland—Bjarni decided to follow his father. The results are recorded in *The Groenlendinga Saga,* one of the great written histories of Norse exploration:

[T]hey put to sea as soon as they were ready and sailed for three days until the land was lost to sight below the horizon. Then the fair wind failed and northerly winds and fog set in, and for many days they had no idea what their course was. After that they saw the sun again and were able to get their bearings; they hoisted sail and after a day's sailing they sighted land.

They discussed amongst themselves what country this might be. Bjarni said he thought it could not be Greenland. The crew asked him if he wanted to land there or not; Bjarni replied, "I think we should sail in close."

They did so, and soon they could see that the country was not mountainous, but was well wooded with low hills. So they put to sea again, leaving land on the port quarter; and after sailing for two days they sighted land once more.

This second place was flat and wooded, and Herjolfsson was even more certain that this was not Greenland, "for there are said to be huge glaciers in Greenland," he told his crew. Even though his men wanted to land for fresh water and firewood, he told them to push on, sailing before a southwest wind for three days to a third landfall.

"This one was high and mountainous, and topped by a glacier. Again they asked Bjarni if he wished to land there, but he replied, 'No, for this country seems to me to be worthless.'"

They sailed along this newest coast far enough to determine that it was an island, then headed off, only to run into a gale. This, at length, brought them to a fourth landfall—and at last it was Greenland.

Bjarni eventually returned to Norway, where he told his tale at court—and was roundly ridiculed for his lack of curiosity and adventure. Other Norsemen were less timid. Leif, son of Greenland colonist Eirik the Red, bought Bjarni's boat and hired a crew of thirty-five, setting out around A.D. 1000 to rediscover the lands from which Bjarni had shrunk.

Leif Eiriksson's journey was Bjarni's, only in reverse. His first landfall he named Helluland, or "Slab Land," where there "was no grass to be seen, and the hinterland was covered with great glaciers, and between the glaciers and the shore the land was like one great slab of rock. It seemed to them a worthless country." This, by most modern reckoning, must have been Baffin Island, the first major landfall west of Greenland. Next they came to a place "flat and wooded, with white sandy beaches wherever they went," which they named Markland, or "Forest Land"—probably Labrador. Finally, after two more days of sailing, they made landfall near a headland, in a shallow bay filled with salmon, "bigger salmon than they had ever seen." They found abundant fodder for animals, and wild grapes, and timber enough to make the journey a financial success. This they named Vinland, or "Wine Land."

Those grapes have posed a thousand-year-old problem for historians. On a course curving west across the Davis Strait and south along the coast of Labrador, it is all but inevitable that the Vikings would have hit the northern tip of Newfoundland—yet wild grapes grow no farther north than New Brunswick, some five hundred miles to the south. Nor do the saga's accounts of a mild, frost-free winter jibe with the reality of Newfoundland. So either the Vikings sailed much farther south than the sagas and common sense indicate, or someone bruised the truth.

The latter seems at least possible. Leif's father Eirik gave the name Greenland to a place made largely of ice and barren rock because, as the saga recounts, "he said that people would be much more tempted to go there if it had an attractive name." It seems likely that the son of the first recorded real estate huckster may have tried the same trick.

What is without doubt is that the Vikings settled in Newfoundland, at the northernmost extreme of the Appalachians. The evidence was uncovered in 1960 by Norwegian explorer and author Helge Ingstad near the tiny fishing village of L'Anse aux Meadows, on an isolated bay at the tip of the Northern Peninsula. Ingstad and his wife had investigated the North American coast from Florida to Labrador over the course of fifteen years, in an unsuccessful search for Viking landing sites; at the time, general opinion considered southern New England as the most likely site for the Vinland landfall. The Ingstads had come to L'Anse aux Meadows with a government medical team, and in response to their questions, a local fisherman showed them the faint, rectangular outlines of buildings beneath the sod along Epaves Bay. The villagers called them the Wigwams, or the Indian Lands, and attributed them to the native Beothuk people. But Ingstad saw a familiar pattern and came back the next year to dig.

Seven years of excavation followed, electrifying the archaeology world with the news that the Norse had settled here, just as the sagas said they had. The remains of eight sod-and-wood houses dating from the eleventh century were uncovered, including an iron forge, supplied with natural "bog iron" from nearby Duck Creek. The houses matched the remains of sod homes found in Norway, and the diggers found nails, stone lamps of Nordic design, even a cloak pin that could be traced not only to Norwegian design but to a particular fashion period. There were objects carved of European oak, and butternuts probably collected on forays south to the Gulf of St.

Lawrence. Near the beach, preserved by burial in the cold, wet ground, wood and bark flakes showed where timber was dressed and trimmed for shipment back home.

L'Anse aux Meadows—the name is a corruption of the French *L'Anse-aux Meduses*, or "Bay of Jellyfish"—is still a tiny village, although today it is accessible by road, and the twenty or so families there have more contact with the outside world than they did in the 1960s. That is due almost entirely to the national historic park that now protects the old Viking settlement, and which attracts some twenty thousand visitors a year—a remarkable number, considering how far this corner of the world is from almost everything else.

I came to L'Anse aux Meadows on a windy, clear day in late summer, when the long grass rippled like the waves on Epaves Bay, and the flowers of the wild irises weaved and bobbed furiously in the wind. Midweek and coming on to evening, there was no one around but the park staff, and I took my solitary time walking down the long path that winds from the bluff to the sea, and to the old colony.

Just beyond the actual settlement, the government has reconstructed three of the sod houses, which rise like sharply peaked hills, as green as the turf from which they were cut; a woven sapling fence surrounds them, and a small longboat rests, keel up, on a rack. If you can ignore the houses and electric wires at the modern village a few hundred yards beyond—if you stand so the sod walls and the longboat's lithe shape frame the sparkling bay—it is easy to imagine yourself in the eleventh century.

Imagining came even easier inside, where a thin stream of woodsmoke rose from the hearth, and the sleeping platforms were a jumble of sheepskin sleeping bags, wooden chests and iron cooking implements. The walls were low and the roof high, rising to a narrow peak of thin, lashed rafters, sodded over with turf; at regular intervals, wooden stakes had been driven down through the sod to hold the layers in place. Square skylights in the roof, ingeniously hinged at the lower edge and propped open with sticks, gave a soft lighting to the interior, showing three long, main rooms and smaller rooms at each end. There was no one around; the park interpreters had gone, and I was alone with my thoughts. Foolish as it now seems, I had a clear sense of trespass, as if I'd walked uninvited into a stranger's home. I left rather more hurriedly than I like to remember.

I sat for a long time in the sun, with my back to the wall of the main house, and looked to sea. There is no way of telling if this was where Leif Eiriksson landed and spent his first winter on Vinland; in fact, it seems likely it was not. L'Anse aux Meadows does not match the description in the sagas, which speaks of a lake connected to the sea by a short river filled with salmon. But there were later voyages to Vinland, and this may have been the result of one of them instead.

The Vikings who came to Vinland were not the horn-helmeted raiders who had been the scourge of Europe from Ireland to the heart of Russia—in fact, some authorities make a point of calling them Norse rather than Vikings for that very reason. These were farmers and traders, lately Christianized (it is thought the reason no graves have been uncovered at L'Anse aux Meadows is that the Norse took the bones of their dead back to Greenland, for burial in consecrated ground). They traveled not in the famous high-prowed "long serpent" ships, but in smaller *knarr* vessels, each about fifty-five feet long and operated strictly by sail. Wide of beam and drawing little water, a *knarr* could be hauled up easily from a shallow bay, but it would carry men, women, children, supplies and livestock through the unpredictable northern waters.

Vikings were not travel dilettantes. When they found a place, they usually stayed—indeed, it is hard to imagine any other European culture successfully colonizing fiery Iceland and glacier-bound Greenland. They were aided in this not only by their tenacity and seafaring skills, but by a bit of good climatic luck as well. The peak of Norse expansion coincided with a period of unusually balmy weather, in which mean temperatures in the northern hemisphere rose about three degrees—a small change on the face of it, but enough to cause marked northward shifts in the tree line and a pronounced warming trend across the northern hemisphere. This change, known as the Medieval Warm Period or Little Climatic Optimum, forced the Inuit cultures on Greenland (and possibly mainland North America) somewhat farther north, providing a relatively empty toehold in Greenland for Viking farms and allowing the transplant of the kind of livestock-based agriculture at which the Norse excelled.

We know relatively little about the Viking colony at L'Anse aux Meadows. It was a true colony, inhabited by women as well as men, as evidenced by the spindle whorl and what are thought to be knitting needles uncovered there. The land would have provided plenty to eat—salmon and caribou, berries, cod and seals from the sea, and right whales just offshore, slow and easy pickings for men used to taking the great mammals. Timber was shipped back home to Greenland and Norway, felled from the great forests of Newfoundland and the Maritimes. That such a long, unwieldy transport could be profitable shows how badly overcut the forests of northern Europe had become.

It has been pointed out that northern Newfoundland was the meeting ground of three great cultures: the Inuit of the Arctic, moving in from the west and north; the Indians of temperate North America, who had come up from the south; and the Europeans crossing the Atlantic. The meeting was not auspicious, at least where the Europeans were concerned. The sagas record the first meeting between the Vikings and a people they called *Skraelings*, a derogatory name meaning "wretches" or "savages."

The Vikings had been living in uneasy (and occasionally violent) co-existence with the Inuit on Greenland, and they were not disposed to look kindly on them. But even that doesn't explain the casual brutality of their first encounter with Natives in Newfoundland, which occurred on the second voyage, led by Leif's brother Thorvald about ten years after the first. Having wintered in his brother's original camp, Thorvald's crew set out the next summer to explore the coast, coming one day upon three humps on the beach.

"When they went closer they found these were three skin boats, with three men under each of them. Thorvald and his men divided forces and captured all of them except one, who escaped in his boat. They killed the other eight and returned to the headland, from which they scanned the surrounding country. They could make out a number of humps farther up the fjord and concluded these were settlements."

The sagas say that Thorvald and his men fell asleep, but that he was awakened by a vision just as a flotilla of skin boats came swarming down upon them. In the fight that followed Thorvald was injured and later died. (Or not, as the case may be—he shows up again in later sagas.)

Dealings with the "Skraelings" soured even further once permanent settlements were established, and eventually, the Vikings "realized ... that although the land was excellent they could never live there in safety or freedom from fear, because of the native inhabitants. So they made ready to leave the place and return home."

No one is quite sure who the "Skraelings" were. The most likely candidates are the Dorset Inuit, for the description of skin boats matches the kayaks of these maritime people. But during the climatic optimum during which the Vikings settled, the Dorset are thought to have migrated from Newfoundland, following the cold north. These may have been the last stragglers.

The other possibility is that they were Beothuk, the Indian tribe occupying Newfoundland during the second phase of European settlement five hundred years later, and which was quickly exterminated. What we know of Beothuk culture does not jibe with Viking tales—they did not use skin boats, for instance, but favored birchbark canoes of great length and unusual design, with a unique raised thwart midway down each side. There is one strong clue, however, that they had met with Europeans before, as set out by Peter Such in *Vanished Peoples*, an account of Newfoundland's native cultures.

> One of the striking things about them, when discovered by Cabot and others, was their extraordinary ability to fashion iron, a trade it seems they could only have learned from the Norse. They must have kept the tradition alive until new sources of iron became available with the second wave of Western Europeans. A surprised Venazano, in 1523, reported that, unlike other tribes he had encountered further south, the Beothuk "had learned the use of iron, so in their exchanges with us demanded knives and weapons of steel."

Whether or not Native attacks were a factor, the Viking colonies did not last long. For L'Anse aux Meadows, the guesses range from a few years to a few decades, but the settlement was soon left to the elements. One reason may have been a sharp deterioration in the climate—the onset of what is known as the Little Ice Age, when average temperatures dropped several degrees. Like the warming at the start of the Viking colonial period, the change seems minor but had far-reaching consequences, provoking cooler summers and much colder winters. More technically known as the Neoboreal, this period of unusual cold reached a peak from 1450 to 1850, but even the first hints of changing climate were enough to tip the balance away from the Norse colonists at the extreme edge of the European sphere.

We know the impact was devastating on the Viking farmsteads on Greenland, the nearest link to the Vinland colonists. Famine followed famine, and the agricultural economy broke down as the Norse refused to change their ways to adapt to the colder climate. The Inuit moved south again, harrying the Norse and wiping out some settlements entirely. After 1400, the remaining Greenlanders were on their own, cut off from Norway by sea ice, just a few years after the last recorded timber-cutting voyage to Vinland in 1397. The Greenland colony was doomed and presumably soon died out.[1]

The interlude between European incursions was brief, but the next prong, when it came, was aimed at the opposite end of the Appalachians. Spanish conquistadors under the command of Hernando de Soto in 1540 were the first Europeans to see the southern Appalachians, but it wasn't by plan. They were lost—indeed, it is hard to be anything but lost when one enters an unknown land with no guides—and they stumbled through the Southeast for four years, leaving their men (de Soto included) in lonely graves along the way. But if they did not find glory, they did find the high hills they named "Appalachian," inexplicably choosing the name of the Appalache Indians in Florida.

The expedition started off well enough. De Soto, who had served with Pizzaro during the sack of Peru, returned to Spain from South America with riches and an eye for the royal governorship of Colombia and Ecuador. He would have been willing to settle for Guatemala, but the Crown had a different idea: title to the Spanish holdings known as Florida, a mostly unknown land north of the Gulf of Mexico. De Soto was to explore this land and, within four years of his arrival there, declare the "two hundred leagues of coast" that would delineate his holdings.

To a veteran of successful conquests in Central and South America, it must have seemed an easy prospect, especially since his expeditionary force consisted of about six hundred men and more than two hundred horses. The Spaniards landed somewhere on the Gulf Coast of Florida and spent the winter, then pushed north.

Almost nothing about de Soto's route can be exactly determined. A government commission in 1939 plotted all the possible routes, based on descriptions written by members of the expedition and others; the potential paths scribble across the map like doodles. The commission's findings have been augmented since then with archaeological data, but we still don't know where de Soto went. "Nowhere can we feel confident that we stand in his footsteps, nor can we even shout through the centuries within his hearing from any one spot. We have no precise hold on his itinerary," wrote Jeffrey P. Brain in the 1985 update of the commission's report.

The expedition headed generally north through central Florida, searching for gold-rich cities, then veered northeast, skirting the eastern slope of the Smokies before turning west through the mountains in May 1540. The logic here is easy to understand, for de Soto knew from his experiences in Peru that great cities often lay in the highlands.

"[T]he way is over very rough and lofty ridges," noted the Portuguese chronicler known only as the Fidalgo [Gentleman] of Elvas, but little was said of the mountains themselves. The expedition found no riches, only small agricultural settlements and increasingly hostile Natives—and no wonder, for this was an undertaking in true conquistador fashion, full of murder, slave-taking and pillage; one Creek "princess" was hauled off as an

The Appalachians formed a barrier to westward settlement, so that for generations after the coastal plain and Piedmont were cleared and farmed, the mountains were still wilderness.

unwilling guide. The Spaniards passed an Indian town whose name they recorded as "Xuala"—possibly near Marion, North Carolina, close to Mount Mitchell and the Black Mountains—then traveled either northwest along the French Broad River or southwest along the Little Tennessee.

Once out of the mountains, their luck only worsened. The constant brutality brought down the wrath of the tribes, and there was a steady attrition of men and horses from war, injuries and illness. The mound-building cultures they met in the Mississippi basin had no caches of gold, but many showed such discipline and ferocity in battle that even the arrogant Spanish noted their courage.

De Soto's band dropped south into Alabama, northwest to the Mississippi River near Clarksdale, Mississippi (where a number of Spanish brass bells, found in a child's grave, provide almost the only archaeological evidence of their passage), then west and south along the Gulf. They were stopped by impenetrable swamps, and it was becoming apparent even to the most ardent conquistador that this new land held none of the gold of Mexico or the Andes. They backtracked to the Mississippi, where de Soto died of a fever—and perhaps of discouragement—in May 1542. His followers tried to take an overland route to Mexico through Texas, but a lack of food forced them back to the Mississippi again. Here they wintered, building seven ships on which they sailed down to the Gulf and west along the coast, finally reaching Spanish settlements in September 1543.

European exploration of the southern Appalachians lapsed for more than a century, even though the English were busily settling the Atlantic coast, and the French were solidifying their control of the Mississippi basin. The mountains, however, held little appeal, and except for a few early exploratory ventures, were largely ignored. In 1669 a young German named John Lederer, acting on the orders of Virginia governor William Berkeley, traveled over the summit of the Blue Ridge in what is now the middle of Shenandoah National Park. It was mid-March, scarcely the ideal time of year for exploring mountains, and Lederer spent six weary days trudging through snow before the cold forced him back.

Before he left, however, Lederer climbed a mountain, which has been identified by some as Hightop, above Swift Run Gap. "The height of this mountain was very extraordinary; for notwithstanding I set out with the first appearance of light, it was late in the evening before I gained the top," he wrote later. To the west he could see the broad valley of the Shenandoah and the rise of the Appalachian Plateau; to the east, he claimed to be able to see the Atlantic—perhaps a trick of light and cloud, perhaps wishful thinking by a man half-frozen.

Settlement trickled in behind Lederer and other early explorers, fueled by the waves of immigrants hitting American shores in the early 1700s. These were valley folk, however, for whom the mountains by and large were a place of unease—a wasteland given over to Satan and savages.

"What did they think of—the first pioneers?" wrote Charlton Ogburn in *The Southern Appalachians.*

... The mountains were a world as far apart for them as those of Antarctica for us today. Farther. The range held the invader blindfolded within its forests and embrasures. Its valleys were choked with rhododendron, its slopes all but barred by the trunks of fallen trees requiring decades to decay, too big even to see over. Most of it was next to impossible to traverse except on trails made by hoofed animals, bears and Indians. Its passes for a man and wife with children, a pack animal with gear and a cow or two were exhausting to surmount. ... Its trees were of such girth that girdling them and waiting for the years to topple the corpses offered to many the only means of clearing a tract for planting. Its depths were unpitying of illness or injury, the peril of Indians unremitting.

And yet they came on, forced by privation in Europe—the Scots and Scots-Irish, the Germans, the Welsh and others. The Appalachians offered the chance of a better life, although who knows how many regretted the choice when they found themselves living in bleak poverty on a far, dangerous frontier?

In 1751, when the Piedmont had been settled and farmed, the Appalachians remained essentially wilderness, traveled by traders, pioneering scientists and soldiers, but few other whites. That year, the Philadelphia botanist John Bartram (father of William) left his home in the company of Conrad Weiser, a skilled negotiator between the colonial government and the Indians. On their way to visit Iroquois leaders on the upper Susquehanna, they rode northwest, crossing into the ridge and valley system just a few miles from my home. Their first obstacle was the ridge the Lenape called *Keekachtetanin*, or "Endless Mountain," and which whites called Kittatinny, or simply the "Blue Mountain."

"... [W]e set forward and ascended the first Blue ridge," Bartram wrote. "The top and south side of this ridge is middling land, half a quarter of a mile broad, and produced some wild grass, abundance of fern, oak and chestnut trees. Descending to the North side we found it more poor, steep and stony, and came soon to the first branch of the Swataro which runs between the ridges ... on this second branch it is good low land, with large trees of 5 leaved white pine, poplar, and white oak, and we dined by a spruce swamp."

Later, crossing the next ridge, "we were warned by a well known alarm to keep our distance from an enraged rattle snake that had put himself in a coiled posture of defense, within a dozen yards of our path, but we punished his rage by striking him dead on the spot."

There is a thrill for any student of history, however much an amateur, to stand in the footsteps of the past. Bartram's directions are fairly clear, and I've retraced his probable path through the hills and along the Swatara Creek, a tributary of the Susquehanna. A hiker still needs to be careful of rattlesnakes, although so many have been "punished" for their rage that they have become quite rare on these mountains.

So have other relics of the past.

"[L]eaving the creek [we] soon crossed another running along the north side of the vale," Bartram continues, "by the bank of which we rode through a grove of white Pine, very lofty and so close, that the Sun could hardly shine through it. ..." There are still pines along the *Keekachtetanin*, but they are striplings compared to those Bartram saw. In pre-Revolutionary days, the best of the huge pines were reserved for the Royal Navy and marked with the ax slashes known as the King's Broad Arrow, but as the hills filled with people, the tall trees fell.

If there is one place that symbolizes the pioneer influx, it is the Cumberland Gap, near the juncture of Kentucky, Virginia and Tennessee. For miles—northeast along the Virginia line, southwest into Tennessee—the Appalachian Plateau rises vertically, walling off the valleys of the Clinch River from the land to the west behind Cumberland Mountain. Even to my modern eye it is an ominous-looking ridge, nearly thirty-four hundred feet high in places and capped with so many rocky outcroppings and cliffs that they look like castle battlements. Only at the Cumberland Gap does the mountain drop its guard, offering a low notch in the hill that permitted relatively easy foot access through the ridges. Even better, the valley of Yellow Creek on the other side leads west through a second gap, The Narrows, in Pine Mountain. The two portals opened up the fertile bluegrass country of Kentucky.

The Cumberland Gap is most closely associated with Daniel Boone, but he did not discover it for the whites. Doctor Thomas Walker and his companions, following game trails laid down by bison, elk and deer (and the Indian war trail known as the Warrior's Path), found the gap in 1750 on an exploration sponsored by a land company. The French and Indian War, however, held settlement in check for the next fifteen years. In 1773, a party of settlers under Boone's direction were attacked by the Shawnee near the gap, but Boone and his workers, paid by land speculator Judge Richard Henderson, cleared the Wilderness Road to the Cumberland Gap two years later. The land rush for Kentucky was on.

Still, the sight of the cliff-scarred mountain, brooding mile after mile above the immigrants as they walked beside their ox-carts and milk cows, must have been a draining psychological experience. Even Boone, the quintessential backwoodsman, felt the menace: "The aspect of these cliffs is so wild and horrid that it is impossible behold them without terror," he said.

While stretches of the Wilderness Road were preserved, the Cumberland Gap for years carried an increasingly heavy burden of modern traffic on a highway laid right over the old wagon road—and below that, the remains of the original game trail. The narrow, twisting road, first paved in 1908, eventually became so congested and dangerous it was dubbed "Massacre Mountain" by locals.

But in a rare case of safety, convenience and historical restoration coming into convergence, the Federal Highway Administration put a new four-lane artery—not through the gap, but under it. By routing the highway through a nearly mile-long tunnel, engineers were able to make the passage between Virginia and Kentucky dramatically safer. And aboveground, the National Park Service consulted old journals and drawings to reconstruct the original contours of the Wilderness Road in 1780, using much of the rock and rubble from the tunnels. Today, a visitor can see very much the same scene that greeted Boone and others heading west into Kentucky.

As the flood tide of settlers of which they were part moved in from the east, the Native Americans they were displacing moved west—sometimes under their own power, sometimes at the point of a gun. The Lenape of New Jersey and Pennsylvania were shoved west to the Ohio and eventually to Oklahoma, and the shuffling of tribal territories up and down the Appalachians had a domino effect on other Native nations. The Shawnee, for instance, were forced north from the Cumberland Valley by the Cherokee, toward the Great Lakes, then east in Pennsylvania, west again toward the Ohio and finally out of the Appalachians altogether.

By and large, the Cherokee had gotten along well with the whites moving into the fringes of their territory. Even though relations with the British were rocky, the Cherokee recognized the value of European trade goods and the shift in the balance of power against their traditional rivals, the Creek. Attakullakulla, the chief whom William Bartram encountered on his travels along the Nantahala in North Carolina, had even traveled to London in his earlier years and had met the king as part of the Cherokee delegation. But during the chief's long life, relations soured; the Cherokee got embroiled in the French and Indian War, first as allies of the British, but eventually falling out with them bloodily in the conflict known as the Cherokee War.

Attakullakulla steadfastly argued for peace, both among his countrymen and the British, but events swept past him. The war culminated with the English, fresh from victory over the French, marching two thousand troops into the Cherokee homeland in 1761 and turning it into a waste of burned villages and destroyed fields. The war ended with the Cherokee scattered, and Attakullakulla and others ceding away title to much of their land.

The Cherokee stood with the British again during the American Revolution, and the difficulties lingered after the colonies became a free nation—even though the Cherokee fought with the Americans against Tecumseh's Creek uprising. (It is ironic, given President Andrew Jackson's shameful treatment of the Cherokee in later years, that Cherokee warriors saved the day for him in the final battle with the Creek.) Meanwhile, the Cherokee had adapted their judicial system and democratic government to the federal model, written a national constitution and invented their own written alphabet, which was used in their own newspaper. They were model citizens in a way many of their white neighbors could not approach.

The tribe's holdings were steadily eroding. The Cherokee executed twenty-one treaties with the British and the Americans, but the settlers kept coming; by 1826, virtually every watershed in their homeland had a few white families on it. When gold was discovered on Cherokee land around 1830, there were increasingly strident calls to remove the Cherokee entirely.

As Michael Frome wrote in *Strangers in High Places*, "It is a truly beautiful parcel of this earth, the Georgia mountains, drained by the Chattahoochee River, which derives its name from the Cherokee word for 'flowing rock,' denoting the many tumbling waterfalls in the Appalachian highlands. It was considered much too good for Indians, better suited to 10,000 gold-fevered men, including many driven to lust and lawlessness, who gorged themselves until lured West by better diggings in California."

Early life in the mountains was hard—hard on the people who lived there, and hard on the land itself, stripped and worn but slowly recovering.

Andrew Jackson, who had been elected president in large measure because of his reputation as an Indian fighter, pushed through a federal law giving him power to seize Cherokee holdings; when the Marshall Supreme Court struck down such "removal laws," Jackson simply ignored them. "John Marshall has made his decision," Jackson is said to have declared, "now let him enforce it."

In 1835 the federal government signed a spurious "treaty" with a handful of Cherokee who acted without the tribe's authorization. In exchange for $5 million, the tribe was to give up its entire land holdings and promise to move to the Indian Territories in Oklahoma. Even though the treaty was a sham, the relocation began in 1838 under the brutal auspices of the U.S. Army, which herded about fourteen thousand Cherokee into concentration camps, then cross-country beginning in October. Predictably, the midwinter trek exacted a horrific toll; nearly four thousand died of exposure, disease and hunger along the way. Little wonder the forced march became known as "the Trail of Tears." (Not all the Cherokee were captured and moved, however. Some hid in the Smokies, living a hardscrabble life in those early years; others managed to return from the west. Many still live on a small reservation just outside the park boundaries.)

With the final barriers removed along with the Indians, the Appalachians came under full white control. The mountains were still the refuge of last resort, however. The cove valleys were fertile enough, but the hills themselves were steep and rocky, the prospect for crops poor and the game quickly shot out. Mountainsides that grew magnificent hardwoods were not meant for corn, and severe erosion was a common problem. The hardships bred a culture forced into self-reliance, supported by hill farms that only grudgingly gave up a living. Horace Kephart, who moved to the southern highlands in 1904 and whose writing often emphasized the privations of mountain life, called it "The Land of Do Without." But one thing the mountaineers had in abundance was a fierce attachment to their rugged landscape.

If the mountaineers were rough on the land, it was unintentional—and the abuses of highland agriculture pale when compared with the rapine inflicted by outsiders. Mineral and timber rights were bought for pennies, often sold by people with little concept of the immense value they were signing away; frequently they were the victims of outright fraud by coal and timber agents. The twin booms of clear-cut logging and coal mining churned across the southern highlands, lining pockets in far cities while providing only temporary employment for the mountain inhabitants. When the booms went bust, they left a land despoiled and a culture tattered.

In the end, tenancy stretching back generations accounted for nothing. Control of huge areas had passed to those to whom it meant only income, and the people who lived in the hills—who collected the chestnuts each autumn, who farmed the coves and hunted the deer—were shoved aside like detritus. Wrote Harry M. Caudill of the mountaineer in his angry book *Night Comes to the Cumberlands*, "Now the trees that shaded him were no longer his property, and he was little more than a trespasser upon the soil beneath his feet." A way of life disappeared on railroad log cars and beneath giant draglines.

Whenever I drive down along the Blue Ridge Parkway through southern Virginia, near the North Carolina border, I make a stop that has become something of a pilgrimage for me. On the slopes of Groundhog Mountain, the park service has preserved the cabin of "Aunt" Orelena Hawks Puckett, who was born in 1837. Married at sixteen, she bore twenty-four children, none of whom survived past infancy—an appalling measure of the pervasive infant mortality in those days. She was skilled at helping other women in childbirth, however, and became the most respected midwife in the area, bringing more than a thousand children into the world—the last in 1939, the year she died at the age of 102.

There's a picture of Aunt Orelena on the historical marker nearby, showing her holding Maxwell Dale Hawks, the last child she delivered. In the photo she is a rigid old woman, ramrod straight with a sunbonnet on her head, sitting in front of the cabin with the baby on her lap. The expression on her face is fierce, but I suspect that was for the camera's sake—being photographed was serious business in those days, and she was remembered by those who knew her as a friendly, kind neighbor.

The cabin is small, with a wood shake roof and two windows, one in the upstairs loft and the other in a downstairs corner. It is empty and locked, so that visitors can only look through the window at the dusty interior. The chimney, once carefully laid, is weathering away, so that now it is caulked only by the clay nests of the mud dauber wasps. The place is redolent of frontier history.

It is easy to forget the toll that people have taken on this land. The scratched, gray photos a century old testify to mountains as worn to the nub as the people they supported. The photo of Aunt Orelena over on the historical marker doesn't show much background, but you can see that the whippy pines and maples around the cabin have gotten much bigger since the 1930s. That's the story up and down the southern Appalachians, where the stripped, logged, degraded hillsides have greened a little more each year. Slowly but surely, the Land of Do Without is healing.

Red-eyed Vireo in American Chestnut Blossoms

Chapter 8

PINUS AND CASTANEA

Castanea

I found a ghost today, flowering in the woods.

It was a chestnut tree, a spindly thing perhaps twenty feet tall and three inches wide at the trunk. Its leaves arched and drooped like green knives, long, finely tapered and edged with sharp serrations. The flowers were starbursts of white, narrow catkins that hung like the sizzling trails of a firework's demise, each strand more than six inches long and furred with tiny blooms.

In human culture, the act of blossoming has become synonymous with life. But for this chestnut, it was all but certainly wasted effort. Although there were other, equally stunted chestnuts scattered through the woods, none were old or healthy enough to flower, and the cascades of blossoms would probably go unpollinated. Worse, on the red-brown trunk of the tree was a gangrenous spot as wide as my thumbprint, the first sign of impending death.

Surrounding the trunk, in fact, were three or four dead stubs about the same thickness as the live sapling, in varying stages of decay. All were chestnuts, and all had risen from the same spot. They were the old hauntings of this ghost, and a portent of what it would soon become.

The American chestnut, *Castanea dentata*, was the crowning glory of the Appalachian hardwoods. A canopy species, it rose to heights of more than one hundred feet, with trunk diameters of four or five feet. Some exceptional specimens were eight feet across at chest-height; others were taped at thirty-four feet in circumference. In the forest a chestnut grew straight and tall; in the open, shading a farmhouse or a stone field wall, it stretched its arms magnificently, as if luxuriating in the elbow room.

Although the American chestnut's original range stretched from the southern coast of Maine to the shores of Lake Erie, and south to western Tennessee and central Mississippi, it was at its heart an Appalachian tree; the core of its homeland encompassed the mountains, plateau and Piedmont from southern New England to the Smokies. Chestnuts grew best in rich, well-drained soil, as high as five thousand feet in the southern Appalachians but on lower slopes elsewhere, usually in association with oaks and hickories, a triumvirate of nut-producers that flooded the forest with mast each autumn.

Chestnut wood was used for everything from cabins to cabinets, fence posts to bed frames, for it was light, easily worked and rot-resistant. As little prone to warping as to decay, it was also prized for floorboards and barrel staves, and more than one wit pointed out that chestnut sheltered a man from cradle to coffin. In late spring and early summer—long after most trees had bloomed—the chestnuts exploded with catkin tassels the color of cream, flowering so thickly that the canopy seemed to have been hit with a snowstorm. And in autumn came the product of that act: the sweet nuts hidden inside spiny burrs, like porcupines concealing a treasure.

That's all done. A fungus no one had ever heard of killed *Castanea* with an almost biblical swiftness and completion, leaving only the roots to send up these brave and pitiable saplings. But the blight bides its time, and before too many years pass the young tree breaks out in cankerous sores that spread and merge, choking it to death. Then the roots, if they have the strength, send up yet another sprout for a few years in the sun.

It is not hard to find a chestnut in the Appalachians. Almost any forest, particularly south of New York, will hold many of the runty, immature trees. It is much rarer to find one that has lived long enough to bloom and rarer still to see a chestnut that has borne nuts.

I found one such tree five years ago, while I was hiking along a portion of the Appalachian Trail near home. I was several hours into the hike and traveling with my eyes down, watching the rocky path with the sort of trancelike weariness that comes with a long trip and a heavy load. The brown burrs did not register at first, but after a few yards I stopped as if slapped, then turned and walked back.

There were dozens of burrs scattered on the ground, some still whole and spherical but most split into quarters. The tree was about twenty feet tall and as thick as my arm, one branch flung over the trail; when I looked around, there were even more burrs to be found.

My first thought was that I'd found an exceptionally large Allegheny chinkapin, a smaller relative of the American chestnut. But the dead leaves had the long, narrow silhouette of a chestnut, not the shorter shape of a chinkapin or a European chestnut. I scratched around in the leaves but found no nuts; it was late in the fall, however, and they may have been carried off by squirrels or eaten by deer—or, one hopes, hidden beneath the leaves, ready to sprout.

Two years later when I went back, the tree was deathly ill, its bark showing the ugly scars of the blight. But it was not dead, and although I have not made the long hike since to check on it, perhaps it has fought off the disease. If history is any gauge, the answer is no.

The chestnut blight took everyone completely by surprise. When it first appeared at the Bronx Zoo in 1904, no one recognized it; it was later surmised that the disease had been accidentally introduced around 1890 with a nursery shipment of Oriental chestnuts, which have a natural tolerance to its effects; another theory implicates a shipment of Chinese lumber. While the introduction of the chestnut blight led to the creation of plant quarantine laws to prevent future tragedies, the damage to the chestnuts was already done.

The blight is a fungus, its microscopic spores carried by wind, bugs, the feet of birds, even droplets of rain. It needs only the tiniest chink in the chestnut's armor to invade—a small crack in the bark, a tiny cut, a bruise. Once inside, the fungus spreads out like an inkblot, its filaments cutting off the flow of water and nutrients through the thin cambium layer as its breaks down the chestnut's tissues; outside, the bark develops a characteristic sunken canker, speckled with small, orange dots called pycnidia that bear the spores. Once the cankers encircle the tree, a process that takes about four years, the chestnut dies, choked by the invader. The spores, meanwhile, have been released to the wind, drifting toward another victim; in wet weather the cankers ooze with sticky tendrils, easily picked up by animals, like the woodpecker from whose feet scientists once washed nearly a billion blight spores. Only the rootstock of the chestnut is unaffected, since the fungus cannot survive beneath soil level.

H. W. Merkel, a forester working in the Bronx, is credited with first noticing the blight, and within a year he was fighting it, doctoring the infected trees and slicing away the cankers. But the blight needed a head start of only those few years, and Merkel's efforts were far too little, far too late. Probably nothing could have stopped it by that point. Because of its ease of transmission, the blight spread with wrenching swiftness up and down the Appalachians, rippling outward like the shock waves of an earthquake from its epicenter. By 1909 it had spread beyond New York City, and by 1915 Connecticut's chestnuts were going fast. New England was stripped of its chestnuts by the 1920s, and as early as 1918, dying chestnuts were found in the Peaks of Otter region of Shenandoah National Park.

The blight moved like lightning, spreading as much as fifty miles per year, despite such desperate measures as a mile-wide "firebreak" in Pennsylvania and programs (including one cold-bloodedly sponsored by a telephone pole company) that encouraged landowners to cut the trees before they died. Heaven only knows what genetic treasure was lost in the frenzy, including, perhaps, some trees with a natural resistance. The blight arrived in the Smokies in the mid-1920s, and by 1938, 85 percent of the park's chestnuts were dead or nearly so. By the 1940s, everything was finished except the grief.

Few people alive today can truly appreciate the sheer magnitude of the loss. I recall my great-grandfather, a timberman in his younger days, speaking with emotion about chestnuts—about going nutting with his family in the fall, about roasting chestnuts for the holidays, about cutting the second-growth chestnuts each winter for as mine timbers, skidding them down the mountains behind teams of mules. His affinity for (and knowledge of) trees was encyclopedic, but chestnuts were obviously something special to him.

In less than a generation, the Appalachians were robbed of their single most important tree—important both ecologically and economically. The toll has been estimated at some 4 billion trees equal to 9 billion acres of forest land, with a value of $400 billion, but even that awesome number does not convey the rending quality of the destruction. A species that made up a quarter of many eastern hardwood forests, and which produced an even greater percentage of its mast crop, had been rendered functionally extinct in an eyeblink.

Interestingly, many authors have glossed over the loss. "[T]he forest closed over Chestnut's place, great as it was, and remained unbroken," Charlton Ogburn wrote, reflecting a general opinion that the Appalachian forests took the blight in stride. It is true that other trees grew up to take the chestnut's physical space, but no other species could fill its niche in the environment. Rather, I suspect, we simply do not understand enough about the intricacies of ecological linkage to comprehend the changes, large and small, forced by the chestnuts' deaths. Or perhaps, like someone who has lost a loved one, we are still denying it all.[1]

It may be, too, that the Appalachians were such a radically altered ecosystem in late nineteenth and early twentieth centuries that the effects of one more insult, however great, were difficult to distinguish from the background destruction. If, for example, the blight had struck a century earlier, when the eastern forests still supported billions of passenger pigeons, it is likely that the fungal assault would have caused catastrophic declines in the pigeon flocks, simply because the birds depended on chestnut mast to such a great extent. While oaks (which moved into preeminence after the chestnut blight) can produce tremendous nut crops, the chestnuts did so predictably, dependably, unlike the sporadic fruiting of oak trees and beeches. As it was, the pigeons were extinct, and mast-dependent big game like wild turkeys and black bears were nearly so—and when these last two recovered, it was with a newfound dependence on acorns.

When the chestnuts died, the most clearly noticeable effects were on people. In southern highlands, the great trees were something of a cottage industry, combining the value of the lumber, the bark—which contained high levels of tannic acid, essential for leather tanning—and nuts. The nuts were the most immediately marketable product, and each autumn families took to the woods to collect them for shipment to the city, where in 1900 a bushel retailed for twelve dollars (the mountaineers, on the other hand, received pennies on the pound for them). A newspaper report from October 1890 tells how "400 sacks and many barrels and boxes" of chestnuts had been shipped from rural West Virginia and Pennsylvania to Wheeling and Pittsburgh by railroad. "The market has already been glutted; only a moderate price can be realized for them," it concluded.

For some mountain families at the turn of the century, the income from chestnuts was important, generating cash as winter was closing in; just as crucial, chestnuts (along with other mast crops) provided a rich diet for hogs, which were driven into the woods to fatten. The loss of the chestnut was a bitter one for some mountain communities.

The mass chestnut death produced some rather macabre situations. With the living supply exhausted, demand for dead chestnut increased, leading to a minor timber boom devoted to harvesting standing, dead wood. (This was possible because chestnut, being so decay resistant, seasoned nicely on the stump, and the posthumous insect damage gave "wormy chestnut" paneling its character.) Ironically, the wood that was once so common and cheap that it was relegated to the most menial of uses—shipping crates, "snake" fences around pastures and similar applications—became sought after, like the work of a recently dead artist that rises in value after the funeral.

Of course, *Castanea* isn't really dead. The tough old roots have continued to send up sprouts, in many cases for more than a century. This tenacity has long given hope to the legions of chestnut supporters that someday, somehow, a cure for the blight will be found, and the monarch restored to its Appalachian throne. Not long after the blight struck, the U.S. Department of Agriculture and others tried hybridizing American chestnuts with Asian species, hoping to transfer a resistance to the blight into a tree that would, at least superficially, resemble the original. They worked on the program for years, but without success. Others tried to plant full-bred Chinese or Japanese chestnuts in the forest, but these much smaller species are, essentially, orchard trees. Even if they had adapted to the Appalachians (which, fortunately, they did not), they could not have replaced the stature of *C. dentata.*

In the 1950s, attention shifted to Europe, where the same blight was having a similarly disastrous effect on the continent's native chestnut species. An Italian plant pathologist named Antonio Biraghi had noticed something odd—some of the blight cultures he grew were white, instead of the normal orange. Even more startling, when some of the white strain was added to a culture dish of the orange, the orange lost its ability to kill chestnut trees.

Biraghi's discovery was met with disbelief at first, but he was eventually proved correct. Somehow, a virus had latched onto the blight fungus, altering it and making less virulent; hence the term "hypovirulent" is applied to this strain. The hypovirulent fungus can still infect the chestnut, but the tree can overcome the disease—and more importantly, the new strain supplants the old, deadlier variety. In the decades since its discovery, hypovirulence has saved many of Europe's chestnut orchards.

An American chestnut infected with the standard fungus can be saved if the hypovirulent strain is carefully introduced. This has proven difficult, however. While the deadly form of the blight moves easily from tree to tree, the hypovirulent fungus does not. In order to treat an infected tree, foresters must pare back the thickest bark to expose the edge of the canker, then use a leatherpunch to make a series of holes around it, into which the altered fungus is plugged.

Obviously, such a treatment program is impractical across the chestnut's entire range, and so far, it has been used only on trees of exceptional value, such as the scattered groves of mature chestnuts growing outside the species' original range, which escaped the blight's first mad rush.[2] The hope was that the hypovirulence would spread naturally among the chestnuts, much as the original blight did (and as hypovirulence has done in Europe), but the weakened strain was maddeningly loathe to cooperate. Nor does it always reproduce true to kind; the spores from a hypovirulent outbreak often revert to deadly strength.

Despite the failures of the USDA to hybridize American and Chinese chestnuts, the private American Chestnut Foundation undertook what became a thirty-year effort to transfer the blight resistance of Asian trees to their North American counterparts.

The technique they use is called "backcrossing," and it is a common method in agriculture for blending the desirable characteristics of two strains of crops. First, Chinese and American trees are crossed, resulting in a generation that is half Asian and half *C. dentata.* Then those hybrids are crossed with American trees, producing chestnuts that are only one-quarter Chinese. Each new generation is crossed with *C. dentata,* eventually swamping all the Asian characteristics except the blight resistance, and leaving a tree that in theory is all but indistinguishable from the original, full-bred American chestnut.

That, of course, is the rub, because as the percentage of Asian genes is pared down, the chances of bumping out the blight resistance also increases—and the only way to know is to put each generation through extensive field trials, which take years. So the backcrossing process, which began in the mid-1980s, has stretched across three decades, though the American Chestnut Foundation is finally working with sixth-generation trees that may soon be ready for planting back in the wild.

Coincidentally, even as the foundation neared its long-awaited goal, scientists may have discovered a much quicker, high-tech solution to the blight problem. Researchers at the State University of New York College of Environmental Science and Forestry in Syracuse have found a way to insert a gene from wheat into the chestnut genome, conferring an immunity to the blight with a molecular snap of the fingers. If the trees gain federal approval (since genetically modified organisms must undergo a barrage of scrutiny) they may be ready to plant around the same time the hybrid backcrosses are—two means to the same, long-sought end.

Chestnut restoration now seems a question of when, not if. And given the dramatic growth of which chestnuts are known to be capable, we could see a transformation of the Appalachian hardwood forests within a single human lifetime—this time not a transformation of loss and destruction, but one of renewal and resurrection.[3]

Pinus

What the chestnut was to the central and southern Appalachians, the white pine was to the northern mountains. Although it grows south to Georgia, the greatest forests—and the greatest pines—were in the north. No other eastern tree could match it for size or sheer grace, nor was any other tree

Echoes of a greater past, a stand of eastern white pines grows in New Hampshire, where some of the tallest pines on record were cut.

its equal as an economic force—in fact, the white pine eclipsed all other species not just in part, but in sum.

But where the chestnut fell before an accident, a blight no one wanted or sought, the great pines fell to greed and calculated destruction. And unlike the chestnut, white pines remain common across their ancestral range. In the century or more since the worst of the timber orgies ended, the pines have soared again—if not so high as in the past, then at least in an echo of old grandeur.

I cannot say why I am so moved each time I see a white pine. It may be that, to my mind, no other tree combines so essentially wilderness a character with such abundance and wide range. When I've been traveling far from home, in alien landscapes where nothing is familiar, the sight that makes me feel most surely that I have returned is the silhouette of a lone white pine's uplifted branches against the sky, or an Appalachian road lined with untold hundreds, straight and solid with their cool, gray trunks, wrapping the hills in feathery, blue-green needles that fill the air with resinous perfume.

Pinus strobus, the great Swedish taxonomist Linnaeus named this tree. The genus name is the classic term for pines (and, obviously, the origin of the modern word), while *strobus* comes from the Greek word *strobo*, a whirling action that refers to the flaring, whorled scales of the cone. It is the only eastern pine with needles clumped in groups of five (one for each letter in the word "white," we were told as children). There is white, if you look closely, two hair-thin lines along the underside of each needle, but the overall color is a cool green, the color you see glowing from the bottom of a deep mountain trout stream. The needles are long and limber, flexible enough so that when you push through a grove of young pines, the result is a tickling caress, not acupuncture.

A white pine seedling grows quickly, given good soil and lots of light. It sinks a strong root and pushes skyward, growing to a bush in a matter of a few years, adding a whorl of horizontal branches with each new season. Leading the way up is "the candle," as the central shoot is called, and under the right growing conditions you'd swear you can actually see the candle stretching to the heavens a little more each day.

In middle age, *strobus* thickens and widens, bulging out at the seams; in an open field, it may expand on this theme for the rest of its life, growing taller and broader with each year. As it ages—as accident and disease broom away branches, as wind shapes the pine with its prevailing cut and the single candle is replaced by multiple, graceful trunks—the white pine takes on an asymmetry that invites the eye to linger. I can look out my window now and see three such trees, growing across the wide opening of my neighbor's cornfield. Their branches carry flattened, upreaching spreads of needles, layered like the scales of a pinecone held upside down or the petals of an unfolding flower bud; through the gaps I can see the many trunks—in truth, old branches that turned to the sky years ago and grew thick. When I stand beneath them and listen to the wind pass through, the needles cut it into a fine, thin whisper, like the hiss of dry snow on old leaves.

In a forest, the white pine's growth is almost all vertical, reaching for the sun. These forest trees don't waste energy on lateral growth; they are tall and straight, with few of the low, spreading branches that make field pines such a delight for children to climb, their hands smeared with turpentinous pine gum. The seedlings may grow thickly (white pine is a masting species, so a heavy fruiting year results in solid age-classes of young trees growing thickly), but they bide their time. If an opening appears in the canopy, from a windfall or timbering, the pines race for the light. Even without such a gap, they eventually reach the canopy, shading out competitors below.

The first written European record of "Pine-trees" dates to 1674, a strangely late year, considering that white pines were quickly recognized as one of the premiere products of the New World. John Cabot, coasting Newfoundland in 1497, reported seeing trees tall enough for masts that may have been white pines, and Giovanni Verrazano, who traced the coast of New England and the Maritimes in 1524, certainly saw the thick pine woods that rolled from the mountains to the sea.

In 1605, English Captain George Weymouth also saw Maine's white pines, and like Cabot—like any good seaman—he recognized their potential as ships' masts. The virgin pines far surpassed anything in Europe, where the best trees had been cut generations before, and the fleets had to make do with masts spliced from several lengths of inferior Scots pine. (Although white pine is known as "Weymouth pine" in England, it was named not for the captain but for the Viscount of Weymouth, who planted its seeds on his estate nearly a century after the captain's discovery.)

Weymouth was right about the white pine's potential for shipbuilding. Its trunks were exceptionally long, light, easily worked and yet strong, and by 1691 the Crown had issued the first of many edicts reserving the cream of the pines for its own use. Any pine, according to a clause in the Bay Colony's charter, growing within three miles of water and "the growth of 24 Inches Diameter and upwards at 12 inches from the Earth" was to be marked with the King's Broad Arrow—an inverted *V* four feet long on either leg, cut into the bark with an ax. At as much as £100 per mast tree, fortunes were made on the tallest pines.

Royal timber agents cruised the river valleys, marking the best trees and infuriating the settlers, who saw the act as an imperial imposition and the trees as legally and morally theirs. In western Massachusetts royal agents marked more than 360 mast trees, but colonists cut all but 37 of them; colonists dressed up as Indians to do the cutting, and London responded with harsh penalties for anyone cutting Broad Arrow trees in disguise. In New Hampshire in particular, sentiment against the Broad Arrow ran so hot that it boiled over in riots. The situation worsened in 1722 when the Crown reserved all white pines, regardless of size, for itself—and retaliated against woodcutters by burning sawmills and flogging miscreants. The trees continued to fall, however, and the seam of anti–Broad Arrow feeling is said to have contributed to New England's hearty acceptance of revolution.

There is a morbid wonder in reading about the old trees; the sizes that were so casually reported make the modern mind—which grows humble in the presence of a hundred-foot pine—reel. One famous pine at Dartmouth, New Hampshire, measured 240 feet high; another, north along the Connecticut River at Lancaster, was said to be 264 feet tall. Yet another, cut for a mast, required fifty-five teams of yoked oxen to drag it out of the forest. From Pennsylvania came a tree that was 12 feet thick and 200 feet tall, and the average size there was nearly 4 feet in diameter and 150 feet high.

White pine has a wide range, from northern Georgia up the mountains to Newfoundland and west through the Great Lakes to Minnesota. It has been pointed out that this range coincides neatly with the region where midsummer temperatures average between sixty-two and seventy-two degrees Fahrenheit, the tree's apparent comfort zone. Not surprisingly, this zone takes altitude into account as well. Found at sea level in the north, white pine climbs to between five hundred and two thousand feet in Pennsylvania, and up to four thousand feet in the southern Appalachians, where it prefers northern slopes and shaded valleys.

The best stands of primal Appalachian white pine extended from north-central Pennsylvania—where it grew so thickly with hemlock that the region was known as the Black Forest—through the Catskills and the Adirondacks, and up the hills of western Connecticut and Massachusetts. But it was in New Hampshire and Maine that *Pinus strobus* reached its greatest glory—and here that the lumber crews wreaked the greatest carnage.

The felling of the Broad Arrow trees was mere prelude. The nineteenth century was the Age of Timber, when population, technology and economy came together to level the old-growth forests. All virgin timber eventually fell: the tuliptrees and chestnuts in the cove forests, the red spruces along the spine of the Smokies and West Virginia's highlands, ancient hemlocks on Pennsylvania's steep-sided mountains, the lowland spruce of Vermont—but the first to go were New England's white pines, which suffered from being too huge and too valuable to be left standing.

Cutting timber was one of the most arduous, dangerous jobs in a time known for the brutality of its labor. Trees skidded from the stump, disemboweling men, or toppled the wrong way, crushing them; limbs (known as widow-makers) fell without warning, splitting skulls. The huge logs would roll off the sleds, and the logjams during the spring river drives would have to be meticulously poled apart by men who might at any instant be tossed into the freezing, crushing mix of wood and water. Little wonder the men who accepted such challenges referred to themselves as Bangor Tigers.

Logging was a year-round occupation, but the fever pitch came in the depths of winter, when the snow lay deep and the thermometer huddled far below zero. Using horse- or oxen-drawn sledges, the men cut and stacked and dragged the logs to the rivers, where they were stockpiled in anticipation of ice-out—those that didn't break free of the sled on the downslopes, mangling horses, oxen and drivers alike.

The river drive was even more hazardous than the winter's cutting—the flood of logs careened down the backwoods rivers on the snowmelt, smashing and jamming into twisted heaps that only a skilled crew with hooked peaveys and guts could untangle. Balance was everything, especially when the jam broke and the logs took to the water like bucking horses. Only a logger's reflexes and spiked boots could save him then—and only if a comrade poling one of the narrow boats known as bateaus was there to pick him up.

There is an inescapable romance in any occupation that, like logging, seems to require a superhuman effort, but the truth is that the loggers lived lives that to modern eyes would barely qualify as civilized. A typical day began hours before sunrise, even in midwinter, and continued until it was too dark to see. Breakfast was beans, rancid pork, flapjacks and gallons of tea; supper was more of the same, along with potatoes and—on the rare occasion when someone killed a moose—some meat. Lunch was brought to the woods to save time for more work. In one of the better camps, the loggers' quarters would be dirt-floored shacks into which were jammed a dozen or more smelly, unwashed men, along with various and sundry vermin (including bumper crops of lice, for which the accepted form of control was to boil one's wool clothing). Bathing simply wasn't done.

The men were understandably proud of their skills, which were considerable. It took incredible stamina to accurately wield a double-bitted ax (one side thin and razor-sharp, the other thicker and heavier), or to know where to set the blade of a crosscut saw so the huge tree would come down safely, without hanging up in its neighbors or smashing to bits when it fell. Average times for an ancient pine: six hundred years to grow, about one hour to drop, delimb and cut into standard sixteen-foot lengths.

By 1840, the white pine along Maine's coastal waterways—the stands most accessible for cutting and transporting—had been cut. Then the timber crews turned to the wild interior, where the pines grew in association with spruces—*strobus* along the rivers and lakes, and wherever the soil was sandy or graveled, and spruce elsewhere.

The Tigers transformed the Maine woods into a trash heap, cutting swath after swath. By the beginning of the twentieth century, nearly three-fourths of Maine's pine forests had gone to the mills, an operation that devastated not only the land, but the water also. The spring log drives scoured the riverbeds, ripping up the substrate year after year; the loss of shade sent summer water temperatures to lethal levels, killing aquatic life—whatever survived, that is, the decaying brew of slash and sawdust that choked the headwaters.

On land, fire became the greatest threat. "Probably by 1900 all of southern Maine that would burn had been burned at least once, perhaps several times, by fires caused principally by settlers and railroads," according to Philip T. Coolidge in *History of the Maine Woods*. While fires sparked by railroad operations were accidental, homesteaders intentionally torched the slash piles, in order to clear away brush for pastures. The result was mile after mile of birch, not a replacement of the original pine.

Some of the fires were terrifyingly big. The infamous Fire of 1825, which raged south of Katahdin, consumed some thirteen hundred square miles of forest; another that same year in New Brunswick, the Miramichi Fire, killed an untold number of people. These and other fires blackened the air with smoke and soot; a distant fire caused a "yellow day," a close one a "dark day." One fire in northern New York in 1780 cast such a pall over New England that birds stopped singing, chickens roosted and people flocked to church, fearing the end of the world. The murk at midday was said to be darker than night.

Once the northern pineries were felled, the loggers looked elsewhere. Many of the Bangor Tigers shipped out west, there to wreak even greater havoc on the white pines of the Great Lakes. The timber barons were looking south, as well, to the stands of *strobus* that grew in the southern Appalachians. It wasn't as clear, light or fine-grained a wood as the northern pine's, but what the hell. "At Shady Valley, Virginia, the yield reached an all-time record, for the South, of 100,000 board feet of White Pine to the acre," Donald Culross Peattie wrote. "So here the industry turned for a last skid to the mills."

The pines withstood this "last skid" for less than fifteen years, ending just as World War I dawned. The railroads inched up the hills, carting down the orangish southern logs—sometimes along the same narrow-gauge tracks that, higher up, were bringing down the last of the virgin red spruce and fir. The Carolina hemlocks toppled on the valley rims, and the Canada hemlock fell in the valleys. The northern hills had been scalped, and now it was the South's turn.

"This aboriginal tree is fast disappearing from the country," John Burroughs warned in the 1880s. "The second growth seems to be a degenerate race, what the carpenters contemptuously call pumpkin pine, on account of its softness. All the large tracts and provinces of the original tree have been invaded and ravished by the lumbermen, so that only isolated bands, and straggling specimens, like the remnants of a defeated and disorganized army, are now found scattered up and down the country. The spring floods on our northern rivers have for decades of years been moving seething walls of pine logs, sweeping down out of the wilderness."

Like Burroughs, you can still find big white pines here and there in the Appalachians, the tag ends that the lumbermen overlooked, or trees that were too inaccessible to be profitable. Whenever I can, I stop to pay my respects at the old groves, which are so few that they have names—the Cathedral Pines in northwestern Connecticut, the Fisher-Scott Pines in Vermont, and others.[4] They are fine trees, many more than 100 feet tall and a few topping out at 150 feet, and they are dramatic to visit—until you stop to think that the biggest of their tribe, the monsters that grew in the most fertile valleys, were nearly twice their size.

And while they are huge and old, they are not virgin. The Fisher-Scott Pines date to the early 1800s, when the first wave of farming in southwest Vermont had peaked and was starting to ebb, allowing *strobus* seedlings to take over reverting fields. Likewise, the famous Catlin Woods pines in Connecticut's Litchfield Hills, which form one of the finest old stands in the Northeast, are themselves second-growth. The area was farmed until the late 1700s, then allowed to revert to timber—and like many pastures, it probably contained several mature pines, left to shade the livestock. These seed trees blanketed the area with young pines (likewise with hemlocks where the ground was somewhat rockier), and after the chestnut blight removed the primary hardwood species, the forest became largely coniferous.

That they have grown so well in less than two centuries speaks to the white pine's inherent vitality, but we're kidding ourselves if we think these scraps are a substitute for the thousands of square miles of truly old, truly enormous pines that once lapped around the Appalachian's roots. Consider the Connecticut pines measured in the eighteenth century, from near this same corner of the Appalachians—trees more than four hundred years old, with a girth of more than fifteen feet. *Those* were giants.

The chestnut, *Castanea*, is an angiosperm, one of evolution's newer twists on plant life. The billows of catkins on a chestnut carry a double bribe—nectar to entice an insect and pollen to be carried by that insect to subsequent trees, thus completing the act of pollination.

Pines are an older lineage, however, one whose origins all but predate the appearance of flying insects. The gymnosperms, as conifers and related plants are known, depend on wind for pollination, gambling on the breeze to drive the engines of reproduction.

The pine bears two kinds of cones—small staminate, or male, cones, each containing dozens of structures known as sporangium that become pollen sacs, and larger female, or pistillate, cones, beneath the scales of which are the egg-bearing ovules. A single staminate cone is capable of producing millions of pollen grains, each consisting of a hard, water-impermeable shell with winglike flanges that catch the wind. If enough pollen is released into the air, some inevitably finds its way to another pine, where it settles down through the scales of the pistillate cone.

At the outer point of the ovule, a sticky secretion catches a few grains of pollen; later, as the secretion dries, it contracts back into the ovule, pulling the grains along with it. A tube forms from within the ovule, drawing the pollen grains farther toward the egg cell shielded within. At last the pollen, which during its transit has germinated and divided into sperm nuclei, fertilizes the egg cell. From such humble beginnings do seeds—and 264-foot white pines—come.

This sort of wind-powered pollination is a risky venture and a hugely expensive one for the pine in terms of energy investment. The amount of pollen released to the air is staggering. The surfaces of many northern lakes, surrounded by forests of spruce or pine, bear a yellow scum in late spring and early summer; this pollen, which may accumulate at a rate of six thousand grains per square centimeter, quickly sinks to the bottom, forming layer after annual layer in the muck—and since the shell of a pollen grain is incredibly resistant to decay, the layers eventually create a fossil record that can be read like a book. Further, each group of plants has its own characteristic pollen shape, so that core samples, bored out of a lake bottom and examined microscopically, give a fairly accurate picture of which plants grew in the vicinity down through history.

Each spring, toward the end of May, I spend a few days fishing a particular trout stream in north-central Pennsylvania. It is a favorite place of mine, and one of the reasons I love this creek (besides the flotillas of mayflies and the trout hanging in the deep holes) are the stands of big white pines that line its banks and range up the steep hillsides. They are second-growth but approaching maturity, some nearly a century old, towering above the chutes and runs of the creek.

As evening comes on and the sun begins to drop, I usually head to a certain pool, where the stream swings hard against the mountain, carving a deep hole that tapers to a wide, knee-deep flat; here, at dusk, the mayfly swarms gather to dance and mate, falling spent and dead to the water. And the trout ascend to meet them, delicately sipping down the insects inside rings of twilight.

Looking downstream from this pool one evening, as the day's wind blew itself out through the ragged crowns of the pines, I could see great clouds billowing against the dark silhouettes of the hills—like mist or dust, but somehow finer and more luminous. The pines had unloaded their stores of pollen to the wind, and a fog of yellow had enveloped the mountains. It swirled and eddied, and when I looked at my clothing, my hands, the lenses of the eyeglasses, I found that I was coated in a talcum wash the color of lemon. I blew a sharp breath against my sleeve, and a small cloud rose and drifted off, mingling with the billions of other grains in the air around me. Of the individual pollen grains aloft that day, only a cosmically small percentage would ever touch the ovule of another pine, but in the process, they would bathe the forest in gold.

White pines are precocious. Trees only five or six years old may begin to bear female cones, although *strobus* doesn't hit its stride until it reaches twenty or thirty feet in height. Oddly, although the number of female pistillate cones remains fairly constant each year—between two hundred and three hundred on an average tree, according to one count—the frequency of pollen-producing staminate cones varies quite a bit from year to year. In years with little pollen production, there will be few fertile seeds; in heavy years, when the small, winged packets are broadcast wholesale to the wind, the ground follows a season later with a flush of green seedlings, springing up like carbon copies. When you see a field of young white pines all precisely the same height, it is probably the result of a bumper seed crop, not someone with a shovel. Generally, the cycle is three to five bad years followed by a good seed crop.

When white pines mast—that is, when they bear seeds in wild profusion—the word gets around. Many of the northern finches feed enthusiastically on pine seeds, among them red and white-winged crossbills, evening and pine grosbeaks, pine siskins and purple finches. These birds also depend on cone crops from spruces, and those years when all the conifers come up dry together, the flocks are forced south in mass irruptions, showing up at bird feeders as far south as the Gulf of Mexico. Some even linger into summer in their newfound homes; after a heavy pine siskin invasion in the late 1980s, birders throughout the mid-Atlantic region discovered the species breeding in backyard conifers, hundreds of miles south of their normal range. By the next season, however, they had vanished—back north, one presumes, to the land of milk, honey and conifer seeds.

Like the oaks and other nut-bearing hardwoods, *strobus* counts on flooding the market, so to speak, with seeds in the expectation that many will escape the seedeaters. That the plan works admirably is plain to see in the abundance of white pines. But the tree has other enemies, against many of which it has not yet developed strong defenses. One of the most common is a weevil, *Pissodes strobi*, which lays its eggs at the tip of the tree's candle. The grub bores down, killing the shoot so that it browns and droops. The branches of the nearest whorl take over as leaders in subsequent years, but they bicker among themselves for primacy, so that the tree loses its stabbing vertical growth and becomes a wider, heavier tree of multiple trunks flaring like a trident's fork, or with a queer bend in the trunk to mark the spot of succession—shapes that make the pine far less valuable for lumber.

White pines must also battle blister rust, a fungus that arrived from Germany at about the same time the chestnut blight was taking off. Like the blight, the rust forms lesions, in the form of blisters, on the branches, trunk or twigs, eventually girdling the tree—particularly young pines, which are especially susceptible to it. There was, it was quickly realized, a prevention. The blister rust is a two-host fungi, alternating with each generation between white pines and gooseberry or currant bushes, so that by clearing away the bushes within about a thousand feet of the pines, the trees could be safeguarded. More easily said than done; Maine spent a whopping $2.6 million on blister rust eradication (mostly grubbing up gooseberries and currants) over thirty-seven years, starting in 1918.

I know a grove of fine white pines in the northern hills of Pennsylvania, the old Black Forest where even the gray squirrels are the color of charcoal, the better to blend with the constant shadows. The trees are so big that when I first came across them years ago, I thought they were virgin growth, missed by the crosscuts and the sawyers. Now I realize they are old second-growth, probably seeded around 1800 when the first cutting began in these mountains.

No matter; they've grown well. I like to sit among their heaved roots, which strangle the mossy rocks, and rest my back against the purple-gray bark of their trunks. On a summer day I can hear the *seet-say seet-say* of a blackburnian warbler somewhere in the green canopy high above, but always there is the wind, whispering through the needles. It has an eager sound that always makes me think the trees are stretching their limbs even higher.

There are seedlings among the giants, little pines only a few feet tall, and they sing the same hissing song of growth. "Time," I whisper back to them. Give it time.

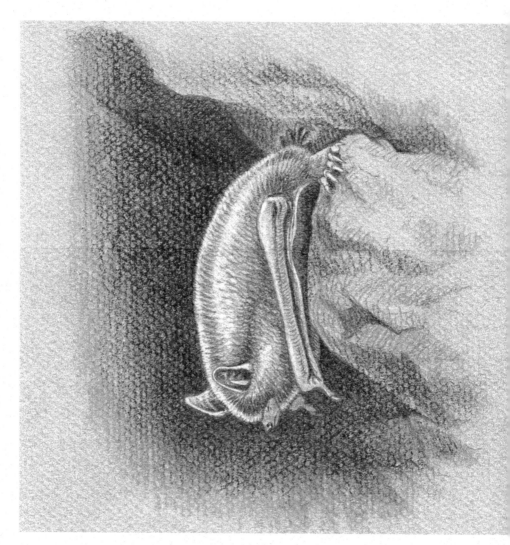

Little Brown Bat

Chapter 9

WINTER IN THE HIGH LANDS

A perfect lens of white amid the rumpled hills that framed it, the beaver pond lay sealed and silent beneath a foot of snow, the trunks of limbless, dead maples standing a gray watch around the edge.

Another Green Mountain winter had settled in for the duration and, save for the hissing of a few icy flakes drifting from the lowering sky, nothing moved.

I'd spent an evening here in the middle of September, just a few short months before. It had been a twilight pressing with life. Little brown bats careened through the quiet air, hawking caddises that danced above the water's surface, and trout, too, rose to the flies. Small flocks of warblers were already on the move, preparing for an evening of migration. A young whitetail snorted indignantly at me from the alders, but I stayed where I was, belly-down on the edge of the dam, peering through the jewelweed at the flat surface of the water.

The sun was still flecking the highest hilltops with light when the beavers finally showed themselves. There was no warning—there never is—only the sudden appearance of a silver *V* as a nose broke the water, then another at the opposite end of the pond. I lay and watched until the light fled, leaving me to strain against the starlight to see their reflections, but not until full dark did I hear the splash and drip that told me the beavers were crawling out to start a night of timbering.

Now, in early January, there was no sign of life, as though winter had robbed the forest of anything animate save the falling snow. The temperature was well below zero, and even within my insulated shell of wool and fiber, I was cold.

It takes no imagination to appreciate winter's threat, and that is especially true in the high country of the northern Appalachians. The season arrives early and stays late, like an unwanted relative that moves in with its baggage of gray skies and endless cold days, and ignores polite hints that it's time to leave. For humans it is a time of snow tires, woodstoves and ski weekends; for animals and plants, the season of reckoning.

Just how severe the northern winter is depends on the year and the place, of course. Some winters are strangely mild, some appallingly harsh. Mount Washington, infamous for its weather, has recorded a twenty-eight-foot snowpack and temperatures as low as minus forty-seven degrees Fahrenheit; the mountain is so high that the jet stream has been known to scrape its summit in winter, bringing gales of more than two hundred miles per hour. The windchills do not bear contemplation.

That is the extreme, of course, but a northern winter even in the more sheltered valleys can be a trial for any living thing, with heavy snow and midwinter temperatures that routinely tumble to ten or fifteen below zero. How plants and animals survive—even thrive—in the face of these conditions is a tutorial in evolutionary genius. For animals, the options are three: flee, sleep or tough it out. For plants, just two: die or live.

Many plants die. Annuals like jewelweed put their faith in their seeds, broadcasting them extravagantly in autumn, then dying completely; each spring brings a whole new generation. Other plants, the biennials and perennials, wither back to the ground, but the roots remain alive, ready to push ahead in spring; most woodland wildflowers, including trilliums, lady's-slipper orchids, hepatica, bloodroot and gentians, fall into this category.

All herbaceous perennials store up nutrients in autumn, not so much to get them through the winter as to give them the energy they need to sprout in spring. Most are content to wait for the strengthening sun of spring to warm the ground, but not skunk cabbage. This arum (the family to which jack-in-the-pulpit, golden club and calla lilies belong) actually creates its own heat, burning carbohydrates stored in its massive root system. The skunk cabbage can raise the temperature of the surrounding soil above freezing, getting an enormous jump on spring. It isn't unusual to find the tops of the greenish-purple hoods poking up through hard-frozen soil in January, and in full bloom a month or two later. When they do bloom, raising hoodlike spathes, the plants crank up the thermostat even higher, keeping the bulbous flower cluster inside at about seventy degrees Fahrenheit—the perfect way to broadcast the skunk cabbage's fetid, dead-meat aroma and attract flies, a main pollinator.

Trees, because of their sheer size, face some special challenges in winter. A woody plant more than a few feet high can't simply die back to the roots each year, or it would lose too much time regrowing to ever amount to anything. And because a tree stands so tall, it is at special risk from the loss of water.

The more immediate threat, however, is freezing. If any tissue—plant or animal—develops ice crystals within its cells, it dies. So if the temperature is going to drop below freezing for any length of time, an organism had better have a way of keeping the ice at bay.

As the air temperature drops steadily through autumn, the tree's internal temperature drops as well, prompting water stored within the cells to migrate out through the cell membrane into intercellular spaces. When this water freezes the ice crystals may poke alarmingly into neighboring cells, but like a finger jabbed into a tough water balloon, the crystals do no harm.

In fact, the reduction of water within the cells serves another purpose—it increases the concentration of the cell's natural chemicals. Just as saltwater freezes at a lower temperature than freshwater, this makes it far less likely that the water within the cells themselves will freeze. Simply by shifting water molecules from one area to another, the tree can prime its tissues with a sort of natural antifreeze. Plants that do not take this trouble, like nonevergreen wildflowers, suffer cellular freezing and tissue death.

These changes take time. If you were to dig up a sugar maple seedling on a hot June day and subject it to freezing temperatures, you'd kill it. By late September, however, the young tree would be able to tolerate a hard frost; several months later, it will be able to withstand nighttime lows of minus forty degrees or worse. The acclimation process, known as hardening, is triggered in late summer by the steady erosion of daylight and accelerates in autumn as temperatures drop. Its stages have been plotted, but exactly what chemical changes bring it on are still unclear.

Forest ecologists have tested many species of Appalachian trees for their lowest tolerable temperature—the killing point at which the tree's defenses fail, and cell death from freezing finally occurs. As you might expect, among many trees—chestnut, several hickories, redbud, yellow birch and others—the killing temperature is close to the normal seasonal low temperature for the northern edge of their range; in other words, they've moved as far north (or as high into the hills) as they can without freezing to death.

But for some reason other trees, mostly northern species, have cold hardiness that far exceeds their needs. Eastern hemlock can tolerate readings of minus seventy degrees, while white and black spruce, tamarack, paper birch, aspen and balsam poplar can all withstand extremes of more than minus eighty degrees. It may be that, like the pronghorns that can still outrun extinct ice age cheetahs, these trees are genetically equipped to handle winters of a severity unseen since the last glacial advance.

Still other trees vary from region to region in cold hardiness. A white pine in the Smokies may be able to withstand lows of minus forty degrees, while one growing in Maine may be able to tolerate readings below minus eighty degrees.

Avoiding frozen cells is only half the battle for a tree in winter. To survive until spring, it must also maintain enough water within its tissues—even though the ground is frozen, and atmospheric humidity levels may drop through the floor. Deciduous trees, with their lightly protected leaves, must shed them in autumn to avoid losing massive quantities of water, but conifers take a different approach. Their leaves are reduced in size but vastly multiplied in number, each needle thickly coated with a waxy surface that seals moisture inside.

In summer, the tiny openings known as stomates, through which the tree "breathes," stand open, but in winter the stomates are sealed like portholes. Still, despite the wax and the closed pores, a tree can lose significant amounts of water, especially if it is exposed above snow level. It was once thought that the snow provided a shield against drying winds, but recent research suggests quite the opposite—that the sort of windy, cloudy days so brutal to animals are actually easier on trees than calm, sunny ones. When the sun is out and the wind still, the conifer's needles heat up more than the air around them. This imbalance increases the rate of water loss to the drier atmosphere. Snow cover reduces this loss, and may even allow the tree to absorb water vapor from the saturated air within the snowpack.

That is not to say that wind is without its hazards for a tree. Anyone who has climbed to the alpine zone of a northern mountain has seen the way the red spruce and balsam fir are twisted and deformed, leaning away from the prevailing wind like children cringing in fright. Wind chews away at their exposed foliage, breaking off twigs and branches and sandblasting all surfaces with abrasive ice crystals. Many trees are "flagged" or "broomed," their trunks denuded on the windward side, with only a few living branches swept back to the lee. In the most exposed sites only *krummholz*, the dwarfed forest, can grow, sheltering beneath the snow as much as possible, and trading horizontal growth for vertical.

Trees are only the most visible forms of life to shelter beneath the snow. Ecologists refer to this as the subnivean environment, where for months on end life chugs forward in a cocoon of ivory, sealed off from extremes of wind and cold.

Snow is the great insulator; poets who refer to a "blanket of white" are more accurate than they know. As snowflakes fall on the ground they jumble randomly, bracing their crystalline arms against each other and creating tiny air spaces; the amount of air varies from a great deal in light, dry snow to considerably less in pelletized sleet. Regardless, this trapped air forms a

In winter the scaly fringes on a ruffed grouse's toes enlarge, like miniature snowshoes, while at dusk the bird itself may fly into a snowdrift, plunging beneath the surface where the snow's insulation will keep it warm through a bitterly cold night.

highly effective insulating layer, protecting the ground from the progressively colder temperatures of the air above and providing a shield for plants and animals below.

For some northern creatures, snow is a temporary refuge. Grouse and ptarmigan are noted for their habit of flying headfirst into snowdrifts, then using their bodies to form small snow caves. Research on ptarmigan shows that such nocturnal burrows may be twenty–five degrees warmer than the open air, allowing the bird to decrease its heat loss by nearly half. Drift-plunging is not risk-free, however; grouse have been killed when they mistook snow-covered rocks for drifts or have been sealed inside by a thick crust of freezing rain.

In many respects, life beneath the snow is an echo of the warmer autumn months. A variety of cold-tolerant arthropods—wolf and jumping spiders, beetles, wasps, mites, millipedes and others—continue to feed and hunt, albeit at greatly reduced intensity when compared to summer's. On warm days it isn't unusual to see the surface of snow take on a gray cast, as though someone had blown fine soot through the forest. A closer look shows that the "dust" is actually minute springtails known as snow fleas, primitive insects that occur by the millions in the leaf litter of the forest, and which come out on mild, bright days as if to sunbathe.

The extent of the subnivean world becomes clear if there is a thaw, and the snow melts. Snow burrows are exposed, a subway system of tunnels dug by mice, voles, lemmings and shrews. On mountains, the tunnels usually fan out from ground burrows or piles of stone, but in meadows or fields, the thaw may reveal globular grass nests built by voles, each the size of a grapefruit, with a round entrance hole on one side. Some of the vole nests may be enlarged and lined with mouse fur; these have been appropriated by weasels, which use the skins of their prey to further pad their new dens. Inside tree cavities (or untended bird boxes) deer mice will build their nests—or they may roof over a bird's nest left from the summer. And many of these small rodents, voles and mice among them, become social in winter, gathering in communal nests to huddle together, sharing their body heat.

Research on the subnivean environment suggests that midwinter is actually a fairly easy time for the small mammals that live beneath the snowpack. Unlike during the warmer months, they are fairly well protected from aerial assault by hawks and owls, although a fox may detect their movements by sound or smell. So, too, may sharp-eared owls; on a number of occasions I've found the wing impressions of great horned or screech owls in deep snow, showing where the birds had plunged for a rodent six or eight inches beneath the surface.

Other hunters take the chase below the snow. Shrews are unstoppable dynamos that require their own weight in food every twenty-four hours just to stave off starvation, and while they feed most heavily on insect larvae and pupae in winter, they will catch and kill mice when they can. A far greater threat to small mammals, shrews included, are weasels, perhaps the ultimate subnivean hunters. Long and thin, they are perfectly suited for squirming along snow tunnels, popping up periodically to periscope the terrain before plunging below, porpoise-like, once again.

The weasel most characteristic of the northern Appalachians is the ermine, or short-tailed weasel, a lithe creature with a body about as big as a chipmunk's, but stretched like taffy and tipped with a bottle-brush tail; the whole animal is only about ten inches long. But forget size. Weasels know nothing of proportion or prudence, and it isn't unheard of for a three-ounce ermine to tackle a two-pound New England cottontail—and kill it.

Like a handful of other northern mammals, the ermine changes in autumn from a brown coat to one of pure white, save for the black tip of its tail. That the pelage change is dictated by length of day rather than temperature has been demonstrated on captive ermine; those kept warm but exposed to diminishing days turned white, while those kept cold but at midsummer day lengths stayed brown. But the change is gradual and the timing varies from individual to individual, so that a lucky hiker might see an all brown ermine and one already white on the same day in early winter.

The black tail tip puzzled early naturalists, who wondered why nature would give the weasel such a perfect disguise and then foil it so thoroughly. The truth is that weasels do not have to hide from their prey (especially since they hunt so often beneath the snow) as much as they must hide from their own enemies. A predator would have to be desperate or foolish to tackle a weasel, but some do, including great horned owls and goshawks. In the heat of a chase, the bobbing spot of black becomes the target, and an owl grabbing for it is most likely to come up empty, or with a few dark hairs for its trouble.

Actually, there is debate as to whether the ermine's white coat is most valuable for its camouflage or for some poorly understood thermal properties. White reflects solar radiation, which would at first thought make it a poor choice for a cold climate until you realize that radiation works both ways, striking the animal from the sun, but also generated by its own body. Perhaps, the reasoning goes, the white fur reflects enough of the weasel's own body heat back again to make it worthwhile to completely molt twice a year; or, similarly, maybe it reflects more sunlight down to skin level than it bounces away. Others argue that since white hairs are somewhat more

A casualty of winter, the skeleton of a deer whitens in the spring sunshine.

hollow than pigment hairs, they offer superior insulation. The question remains open, and the ermine, completely oblivious to the discussion, keeps changing with the chill and thaw of the seasons like it always has.

While small mammals are able to gain insulation and concealment from snow, there are difficulties. Carbon dioxide levels can rise to uncomfortable (even toxic) levels when snow conditions and topography conspire to trap the gas, forcing animals away from the concentrated pockets. A prolonged warm spell that melts the upper layers of snow can also cause flooding—and if the warmth is followed by a cold snap, the melted layers may refreeze in a solid crust, limiting the mobility of animals like weasels that switch between the surface and subnivean zone.

For plants, the world beneath the snowpack also offers challenges and advantages. Lacking the conifer needle's waxy cuticle, herbaceous plants like perennial wildflowers would quickly die from desiccation if exposed to winter winds, but beneath the snow, surrounded by relatively humid air, many are able to remain green and hydrated all winter. If the snow is shallow enough to allow significant amounts of light to penetrate, the plant may even be able to photosynthesize in a limited fashion. Even so, when spring comes and the snow melts, the plants have a head start over those that must send up new growth from a root system.

Not every living thing in the high lands can shelter beneath snow, of course. Small birds like finches, kinglets and chickadees have an especially difficult hurdle to overcome. Because they are so small, they have a great deal more surface area (which loses heat) in relation to body mass (which produces heat).[1] To compensate, most cold-climate birds grow more feathers in autumn, in some cases increasing feather weight 50 to 70 percent over summer levels, and can fluff them using tiny muscles at the base of each feather shaft, increasing the amount of trapped air. But mostly they shiver.

Songbirds shiver, in fact, almost nonstop through the winter. The muscle tremors generate heat (the same reason we shiver when we become chilled), keeping the bird's body temperature stable—but at a price. Heat production requires fuel, so that small birds must feed incessantly, and because they must keep their weight low enough to fly, they can store up only enough fat for one or two days without food. By one estimate, a crossbill must eat an average of one spruce kernel every seven seconds, sunup to sundown, just to break even metabolically. No wonder the chickadees at my feeder seem infuriated when I'm late filling the tubes with sunflower seeds.

The shivering response kicks in at different temperatures for different species and, as might be expected, birds with more southern affinities have less tolerance than northerners. Cardinals, which only moved into New England within the past century, begin to shiver when the air temperature drops to just sixty-four degrees Fahrenheit. Blue jays shiver at the same point, while their boreal cousins, the gray jays, do not shiver until the thermometer drops another twenty degrees.

Chickadees are also known to drop into a light torpor on cold nights, lowering their body temperature by as much as twenty degrees and reducing their energy demands by about a quarter. A torpid bird is, of course, unable to rouse instantly to flee a predator—but by restricting torpor to cold nights when it is in a safe roost already, the chickadee can afford to let down its guard a bit.

A moose is at the other end of the thermodynamic chart from a chickadee. Weighing more than half a ton, a bull has a healthy balance between surface area and mass, and its heavy coat of densely packed, hollow hairs gives it superior insulation. Indeed, a moose can quickly become overheated on a calm, sunny day, even if the temperature is near zero.

A moose's ungainly build—a heavy body perched on ludicrously tall, thin legs—pays off during a northern winter. The long legs are designed to rise above the deep snow, allowing the moose to move easily even through heavy drifts. Caribou in the Long Range Mountains and the Shickshocks are even better suited to snow than moose. Researchers E. S. Telfer and J. P. Kelsall compared the physical adaptations and behavioral traits of a number of wild grazing animals to create a "snow-coping index," and caribou came out on top by a wide margin. Their feet are enormous, spreading the weight over a wide area (they also have extra-long dewclaws splaying out behind for additional support), and their legs are long, like a moose's, allowing caribou to step easily through heavy snow. In addition, caribou adapt behaviorally to winter—migrating from areas with the heaviest snowfall, making single-file trails to minimize work, and digging efficiently through drifts for food.

Deer, on the other hand, lack the stature and limb flexibility of moose or caribou, and so become bogged down in even moderate snow. In consequence, whitetails in northern forests must gather in sheltered spots (often conifer stands) known as yards, where the snow becomes trampled and easier to move around in. Unfortunately, a deer yard doesn't offer much in the way of nourishment, and mass starvation may result if the period of deep snow goes on for too long.

Once in central Maine, walking along the shores of a small lake in early summer, I came across the carcasses of five or six deer, white bones and old skin beneath the dark needles of a spruce stand. The skeletons were disarticulatcd, worried apart by scavengers, the heavy hair lying in thick, ragged mats around them, like gray halos. There was hair, too, in the whitened piles of coyote scat left here and there through the yard, mute evidence of where the deer wound up. There was a terrible poignancy in the scene, however natural this outcome, and the sense of melancholy lifted only on my hike back. Stepping over a log, I surprised a newborn fawn, which bolted unsteadily away through the forest. I put my hand in the place where it had lain and felt the warmth of the leaves through my palm.

Snow and cold are of little consequence to a bear, which may face winter with the greatest nonchalance of any northern Appalachian mammal. All autumn the bear is little more than a mobile food-processing station, converting acorns, beechnuts, apples, wild grapes, carrion, field crops—in short, almost anything—into fat. The weight gains are eye-popping; a male weighing 300 pounds in midsummer may balloon to 450 by the time November arrives, adding weight at a rate of 2 or 3 pounds a day. On some bears, the fat layer beneath the skin may be four or five inches thick.

Depending on the weather and the local food supply, bears begin looking for a den sometime in late autumn or early winter, with pregnant females turning in earliest, and young males as much as a month later. Caves, which most people assume are the preferred winter quarters, are usually avoided, although bears will often den inside deep rock piles or between boulders—they like tight accommodations.

Black bears are as flexible in their choice of dens as they are in their diet; they have been found in hollow logs (both fallen and standing), timber slash piles and natural blowdowns, highway culverts (where sudden flooding may drown them), and beneath the porches and decks of backwoods cabins. In the more temperate southern Appalachians, they may spend the winter in tree dens, including open nests. While bears are normally solitary in winter, a female may den up with her yearling cubs, packed together like so many black, furry sardines in a space you would scarcely credit with holding one bear, much less four. And occasionally a bear will simply find a slightly sheltered hollow and curl up on the ground, utterly indifferent to the elements.

Some bears take a more sybaritic approach. In Pennsylvania some years back, a 290-pound bear broke a small window into the basement of a home, scooped bushels of leaves inside to make a nest next to the hot water heater, then pulled most of the insulation down from the floor above. In the process, he also wrenched loose the vent from the clothes dryer upstairs—so that all winter long, whenever someone did a load of laundry, he was bathed with hot air.

Such excesses aside, people rarely know when there is a bear denning in the neighborhood. Invariably, when biologists follow a radio-collared bear to its den beneath a backwoods home, the human residents have no clue that a four-hundred-pound carnivore is curled up beneath their feet. Most, understandably, react badly at first, but there is no reason for panic. The bear will sleep quietly through the winter, then disappear without a sound come spring.

Black bears are not true hibernators; they do not experience the profound, almost deathlike torpor of a bat, woodchuck or jumping mouse, whose metabolism drops to a fraction of its summertime level. Bears are winter dozers, groggy but still able to wake in a hurry. Biologists like my friend Gary Alt, who crawl into winter dens to dart sleeping bears, know this well. Gary has been run over on a few occasions by "hibernating" bears that came instantly awake at the jab of the needle.

With winter imminent, and having gorged itself to a state of obesity, the bear makes one final preparation before denning—it eats an ungodly mess of twigs, grass, pine needles and stems that form a solid fecal plug, stopping up its digestive system like a cork.[2] Other changes are physiological rather than mechanical; the bear's temperature drops about ten degrees and respiration and heart rate decline marginally, while its metabolic rate (the rate at which it consumes energy) is cut in half. Its system shuts down for five or six months, and it will not eat or drink through that time, subsisting entirely on its fat reserves. Exactly how it manages to metabolize stored fat (it will lose up to 25 percent of its weight by spring) without being able to rid itself toxins by urinating is of great interest to medical researchers, who imagine applications for humans with kidney disease.

True hibernators take this whole process to the extreme, paring away the signs of life until only the barest flicker remains—just enough to keep the spark going until spring. There are fewer hibernating mammals in the Appalachians than people think: bats (those species that do not migrate), woodchucks and two species of jumping mice. Chipmunks, which with bears are often considered hibernators, do drop into fairly short periods of torpor, but because they lay on no heavy supply of fat, they must wake periodically through the winter to feed. This they accomplish by visiting their underground food caches, which may hold up to a bushel of seeds and nuts, but if the weather topside is moderate, they may even spend a day or two actively foraging on the surface.

Bats are the quintessential hibernators, and because their hibernacula (as hibernation sites are known) are fairly accessible, science knows a fair bit about how they manage the feat. Not all bats hibernate, however;

three species known as tree bats, distinguished by their thickly furred tail membranes, migrate south from the northern Appalachians each autumn, sometimes traveling as far as Bermuda, Mexico and Central America. But the half-dozen or so other species found north of New York are hibernators, although they, too, may travel long distances to reach their hibernacula.

Some bats (which can be banded like birds to track their wanderings) have been known to travel hundreds of miles between summer territories and wintering sites—an indication, perhaps, of the difficulty in locating a hibernacula with the precise conditions a bat needs if it is to survive the winter. Caves and old mine shafts are most often chosen, as long as they have a stable temperature of between thirty-four and forty degrees, with saturated humidity; some species, like big brown bats, can tolerate colder, drier conditions, sometimes wintering in attics and within uninsulated walls.

Bats walk a metabolic tightrope in winter. Because they are flying animals, they (like birds) are limited as to the amount of fat they can store up. Nevertheless, a little brown bat going to hibernation may carry as much as a quarter of its body weight in fat, the result of countless evening forays for insects. The bats choose the region of the cave where they instinctively sense the proper balance of temperature and humidity, clustering together in crevices or blanketing the ceiling. Their body temperature drops to roughly that of the surrounding air, and respiration and heart rate plummet as well. Bathed in the cold damp of the cave air, their fur may even take on a netting of condensation, so that they sparkle like gems in the beam of a flashlight.

Unfortunately, a human visitor may doom the bats. It is hideously expensive, in terms of energy, for a hibernating mammal to be roused from a deep torpor. Bats simply don't have the reserves to permit repeated disturbances, and cavers who wake a colony may sentence them to starvation before spring arrives. For this reason, many of the better-known bat hibernacula are gated against entry by anyone except biologists—at least during winter. Because the caves aren't used by bats in summer, they can be safely enjoyed by humans then.

Land animals that we are, we think mostly in terrestrial terms, but winter brings changes below the water's surface, too. As early as October in the higher elevations, the nocturnal ice begins to form, a thin moonbeam skin that creeps out from the shoreline and disappears with the sun. Although the lakes and ponds hold tremendous quantities of heat stored within their water, they are steadily cooling through autumn, bleeding that heat away into the atmosphere. The stratification of the water into warm, light layers over cooler, denser ones, which was the norm all summer, breaks down, and during the period known as autumn turnover, the entire lake circulates with the wind, mixing oxygen-rich surface water with gas-poor water from the bottom. It is a final gift from the atmosphere before the imprisonment begins.

With the first hard freezes of winter come nights of bitter, windless cold, when the skim ice deepens to a heavy sheet, so clear that the darkness of the water shows through. The sun, when it rises, is too weak to break its hold, and the cap continues to grow. This is the "black ice" of early winter, beloved of skaters, a windowpane over the lake that later thickens into "milk ice," cloudy with trapped air and gases, buckled and cracked and groaning with expansion.

Locked away from the air, the waters of the lake still support life, of course. Fish in particular are little affected by the change, although the cold may send some, like bullheads, into a near-torpor. Others, like sunfish, continue their lives at a reduced pace, their metabolisms scaled back and their activities a lazy shadow of summer's. Still others that thrive in cold water, such as trout, are almost as active as ever.

As the ice turns from transparent to translucent and the snow piles up in ever-deeper drifts, the light penetrating to the water diminishes, as if a giant curtain has been drawn across the sun. The lake is wrapped in murky darkness, and the fish swim in a twenty-four-hour-a-day night. Yet the plants of the lake—both the microscopic phytoplankton that drift free in the water and the rooted plants that blanket the shallows—continue to use the dim light for photosynthesis, manufacturing oxygen as a by-product of producing food.

This is critical, because the amount of dissolved oxygen, not temperature, is the single most important variable in the winter lake. The ice forms an effective barrier to gas exchange, and the microbes that populate the bottom sediments guzzle the precious stock of oxygen as winter progresses. Submerged plants cannot make up the shortfall—in fact, at night they must also consume oxygen, adding further to the erosion of the gas—but their daytime work at least softens the blow to other organisms.

In a sense, the animals and plants of the northern lakes are in a race with winter. If the ice lingers too long, or if (as in a shallow pond with organically rich sediments) the rate of decomposition is unusually high, the oxygen may simply give out. A few years ago the winter was long and hard, with deep snow cover late into spring that kept the water dark, suppressing the surge of photosynthesis that normally comes before the thaw. When the ice finally melted on many ponds, it revealed the white, pasty bodies of hundreds of fish, suffocated in the waning days of winter.

Given the risk of oxygen depletion, it would be helpful if an organism could just do without the gas. Remarkably, at least one resident of the winter lake has figured out how, along with a suite of behavioral and physiological adaptations to winter's cold.

On a bitterly cold night many years ago, I was shining a flashlight through two inches of black ice on a neighbor's pond when something moved below the surface. I thought it must have been a muskrat, but a moment later it rose again, and I could see that it was a painted turtle, swimming with exaggerated slowness beneath the ice.

I'd known that painted turtles, like many species, wintered in the mud of pond and lake bottoms, and I knew that they were always the first reptiles active in spring, basking when the lakes were still half ice-covered. I'd never realized, though, that they could exhibit even this limited degree of mobility in the middle of winter.

In fact, painted turtles are the most cold-tolerant of North American reptiles. Although they emerge from their eggs in early autumn, baby painted turtles remain in the underground nest over the winter, even though they are likely, to human eyes, to freeze "solid"; all physical processes cease except for minimal brain activity. But just as trees avoid damage at low temperatures by carefully segregating ice crystals within their tissues, so too do young turtles survive this harsh treatment by avoiding cellular freezing. Unlike trees, however, turtles actually produce chemicals that keep ice crystals small and manageable, while at the same time boosting sugar concentrations within the cells to lower the freezing point.[3]

Once they reach adulthood, this ability to survive a solid freeze disappears, and painted turtles that hibernate in a shallow area prone to ice may die. Consequently, turtles are careful to choose a spot deep enough to avoid bottom-fast ice, and one that is relatively warm.

Warm? When we stick our hand in winter water, it feels instantly frigid—far colder than the air, even on a below-zero day. But this is a trick our warm-blooded bodies play on our minds. The water is substantially warmer—at least thirty-two degrees Fahrenheit, or else it would be frozen. It is only because water conducts heat so much more efficiently than air that it feels colder to our hands. For a cold-blooded animal like a turtle, whose body temperature matches the surrounding lake water, there is an advantage to hibernating in bottom sediments where the temperature may be ten degrees above the freezing point.

Oxygen deprivation is the greater worry. Cold water can hold a great deal of dissolved oxygen, which a hibernating turtle can absorb through the lining of its throat skin, but as the late-winter oxygen crunch comes on, the turtle—like all aquatic life—faces a crisis. But where fish will die if oxygen levels decline too far, painted turtles are able to withstand an almost complete absence of oxygen; in fact, they may invite the problem by digging down into oxygen-poor sediments. Researchers who have studied wintering turtles have found that they can live in oxygen-free water for almost five months, burning stored carbohydrates for fuel and buffering the toxic acid by-products with calcium drawn from their shells.

And what of the beavers, whose dammings have created so many still-water habitats in the northern Appalachians? Warm-blooded, air-breathing mammals, they face some of the most difficult hurdles in surviving beneath the ice. They do not hibernate, and unlike turtles they cannot go for more than a few minutes without a breath.

Part of a beaver's solution to winter is structural. The lodge, which is integral to the dam-and-pond complex, rises head-high above the water's surface, a conical mound of sticks and logs. Hidden beneath the surface are one or more entrance tunnels that lead up to a central chamber above water level, where the beavers—adults and nearly grown kits—can huddle and groom. The peak of the lodge is made up of somewhat more loosely packed sticks, but the whole mass is solidly cemented together with mud, now frozen to the impregnability of a castle. It is safe not only from predators but from the cold; trapped body heat may raise the interior temperature to a toasty forty-five or fifty degrees Fahrenheit when the outside air is below zero. When they're swimming, their thick pelage of long guard hairs over a dense undercoat seals in air next to the skin, so the beaver's body stays dry and warm.

The beavers, true to the cliché, are busy all summer and autumn, storing up food for the winter—because they do not drop into hibernation, their metabolic requirements are largely undimmed. The larder is a cache of tree branches and limbs, jammed into the muddy bottom near the lodge and kept fresh by the natural refrigeration of icy water; the nutrient-rich inner bark is stripped off and eaten as needed. The beavers need not ever step out of the water from ice-up to ice-out.

Still, they must breathe. Most forays, including those to the larder, are accomplished with a deep breath taken inside the lodge. A beaver can make the most of that breath, holding it for up to fifteen minutes on dives that can stretch for over half a mile, although most dives last only three or four minutes. But if pressed for air—and I find this one of the most remarkable feats in the winter mountains—a beaver may actually rise to the top, pressing its flat nostrils against the bottom of the ice, where a thin layer of air may be trapped. This is often the result of old exhalations by the beavers themselves, used air in a way, but because of gas exchange it becomes recharged with oxygen. Carefully the beaver inhales, then plunges deep again, refreshed.

Walking on this frozen beaver pond in the Vermont mountains, I try to picture that trick from the beaver's perspective. I cannot; the horror of being trapped beneath the ice with my lungs running out of air gives me a chilling, momentary sense of panic, even standing beneath the open sky. Somewhere below me, the beavers are probably swimming, in a world so completely alien that I cannot imagine it.

The snow had started as I dropped east out of the Adirondacks, and by the time I crossed Lake Champlain into Vermont at Crown Point, the visibility had dropped to almost nothing in squalls. It was just after New Year's, the thermometer was having a post-holiday sulk somewhere near zero, and the Green Mountains were lost in gray clouds.

The road crossed Dead Creek, the big wildlife management area that sits in the flat Champlain Valley, and here the snow eased up for a moment. I could see the skeletal elms and the gray silhouettes of wooden fence posts marching away into the white, each post capped with a hat of snow. One cap looked odd, somehow, and I slowed even further, peering through the flakes. With a start, I realized that what I'd taken for a lump of snow was actually a huge snowy owl, pale as a drift, staring back at me with luminous yellow-orange eyes.

Snowy owls are Arctic birds that spend virtually all of their time along the tundra coastal plain that borders the Arctic Ocean, and on the high-latitude islands like Baffin and Ellesmere. They are not migratory in the normal sense, but a few come south each winter—and when their prey (primarily lemmings) experiences a population boom, snowy owls from across the Arctic congregate to feed on the bounty, and to produce a bumper crop of chicks. That winter, legions of young snowy owls will pour south in a phenomenon known as an irruption. Many hug the coast or the Great Lakes where the horizons are as flat and open as their northern homes,

Tracks tell the tale of a cottontail ambushed by a great horned owl, which left its wing prints in the snow around the picked carcass.

but a few come down the Appalachians, fetching up in farmland like the Champlain Valley.

The owl and I watched each other for less than a minute before caution got the better of me; a snowy road is not the best place for birding. I left it sitting placidly in the squall, as even thicker flakes began to swirl.

There are a number of other raptors that take a similar approach to winter, budging only when food is low—goshawks, boreal owls, great gray owls and northern hawk-owls. Also irruptive are the northern finches that rely on the unpredictable cone crops of spruce and pine—the grosbeaks, siskins, crossbills, red polls and purple finches.

In a good finch year, bird feeders from Maine to the Carolinas will be brimming with northern visitors, and feed mills report runs on sunflower seeds. Evening grosbeaks in particular can consume gluttonous quantities of seed; one friend used to refer to them as "gross-pigs" for both their table manners and their appetites. During an especially good grosbeak winter, another friend went through four fifty-pound bags of sunflower seeds each week from November to early April—nearly two tons of seed, most of which disappeared down the gullets of the hundreds of grosbeaks that turned his feeding station into a yellow and black wonderland.[4]

Irruptive raptors are much more secretive, and when one is found, the word gets around birding circles quickly. Many years ago someone discovered a northern hawk-owl—a peculiar, day-hunting owl of the Arctic that resembles a falcon—along a dirt road in the Poconos. It was the first time in nearly a century that the species had been found in Pennsylvania, and by the following weekend, news of the owl had flashed across bird-alert hotlines up and down the East Coast. Hundreds of people made the trek to see the bird, to the utter bemusement of the two or three farm families living nearby. One of the residents told me, with some wonder in her voice, that a couple had come all the way from Arizona to see it.

I heard about the owl the day after it was discovered and managed to beat the crowds. I made a stop at a pet store for some mice and hit the road for the three-hour drive. The owl was sitting just where it was supposed to be, near the top of a clump of leafless aspens; it looked so much like a kestrel, with a long tail and big, chunky head, that I very nearly overlooked it.

Arctic owls are famous for their lack of fear; most have never seen a human before coming south and treat us as just part of the scenery. I pulled off the dirt road about a hundred yards from the owl and stepped out, carrying a plastic terrarium with a half-dozen mice in my hand. The hawk-owl snapped its head around, bobbed it up and down a few times to triangulate the distance, then launched itself straight at me. Even in flight it looked more like a hawk than an owl, with long, falconish wings beating hard.

I stood motionless, holding the box of mice. The owl arrowed in closer and closer, never once looking at my face. It flared up right in front of me, whacking both feet against the plastic; only then did it glower at me with an accusatory look before landing on a fencepost a few yards away.

I opened the lid and set a mouse on the ground at my feet. Within seconds the hawk-owl had dropped and made the kill.

Over the next two days, I spent hours watching and photographing the hawk-owl. I doled out the rest of the mice, but most of the time I was content to watch it hunt meadow voles in the alfalfa field that bordered the road. It ate a few, but most of the rodents were stashed away—some in dense grass along a bank, others in a cavity in a dead oak.

Such caching of food is common among many northern raptors, especially during the breeding season. In winter, when low temperatures preserve food for long periods, the value is obvious, but there are drawbacks. Eating icy meat could fatally lower a bird's internal temperature, so many owls will "incubate" a frozen mouse under their breast feathers until it thaws, and then eat it. Twice I saw the hawk-owl recover stiff frozen voles from its ground cache and sit on them for more than an hour before eating.[5]

Some owls are irruptive and some, we are coming to understand, are regularly migratory. Not so the great horned owl, which generally hunkers down for the winter on its own turf and toughs out the season. Among the most powerful owls in North America (the great gray is larger but is mostly fluff and feather), the great horned owl weighs about four pounds and has a wingspan of up to five feet. Its strength is awesome and a little frightening; by some accounts it can exert two thousand pounds of pressure per square inch with its feet, powering eight needle-sharp talons. I know a biologist whose captive great horned once unintentionally crushed—not merely broke, but crushed—the bones of her hand when it clenched in fright while riding on her glove. Great horneds are capable of killing animals as large as adult snowshoe hares and Canada geese. No wonder winter doesn't faze them.

In fact, the great horned owl does more than simply wait out the season—it turns convention on its head by breeding during the dregs of winter. By the December solstice the pairs have begun to court, calling back and forth through the frozen night, the male smaller in size but deeper-voiced, the big female with a higher-pitched call. They lob their song, a string of five- to seven-note hoots, back and forth through the calm, still air. They will also react quickly to the calls of intruding great horned owls within their territory, a trait birders exploit when doing a census. Once, while helping my friend Kerry tally owls in the predawn of a Christmas Bird Count, I watched a great horned pass within inches of his head while he played a tape recording of its call. Only at the last moment did the owl realize that Kerry's brown wool cap wasn't a trespasser to be knocked silly.

Great horned owls do not build their own nests; instead, they appropriate the old efforts of crows, hawks or squirrels, or choose the hollow cavity of a snapped-off tree. The eggs—usually one or two—are laid in February across much of the northern Appalachians, a time of especially heavy snow and damp cold. That they survive is testament to the zealous attention of the parents, particularly the female, which broods them through snowstorms, freezes and sleet. Within a month the chicks hatch, confronting at least as hostile an environment as they faced as eggs. Here, too, parental brooding keeps them alive, although the young owls are cloaked from birth in a dense coat of down and can stand exposure to the cold within a few weeks.

A good thing, too, because like many owls, young great horneds have a peculiar habit of jumping ship early in the nestling phase, weeks and weeks before they can fly. Some merely scramble out of the nest, scattering through the limbs of the nest tree; others tumble clear to the ground, bouncing harmlessly. In most bird species this would be a death sentence, but the "jumpers," as they're known, have a survival rate almost as high as those that stay put in the nest. Credit the parents again; unlike many birds, which do not seem to recognize their nestlings once they are removed from the nest, great horned adults will find and feed their wayward chicks and defend them with hard, slashing attacks should predators come too close.

If such a precocious nesting season seems maladaptive, consider that it takes more than four months from egg-laying for a young great horned owl to even begin to master the skills needed to hunt for itself. If owls followed the same timetable as, say, robins, with courtship and egg-laying in April, the young would just be reaching self-sufficiency as autumn dawned. This way the owlets have a long summer of practice, filled with the incautious young of other, Johnny-come-lately birds and mammals.

Few mammals are able to follow the owl's example and breed in the winter, although some small rodents like voles, protected beneath the snowpack, may bear a litter or two. Only the black bear makes a habit of birth in the coldest months, all the more miraculous when one considers that the mother is asleep most of the time.

Black bears mate in midsummer. Unlike many wild mammals, among which the female comes into estrus for only a brief period, black bears stay in heat for weeks, until they finally mate. And only during the breeding season do these otherwise solitary animals socialize, a noisy, ribald courtship that may include spectacular fights between rival males, but which more often involves surprisingly tender moments of nuzzling and stroking between the male and female. They stay together for a period of days or even weeks, mating a number of times before her interest wanes and he goes off in search of another receptive female.

But just after impregnation, an odd thing happens. The fertilized egg, which would under ordinary circumstances adhere to the wall of the uterus and begin growing, does not. It undergoes several cell divisions, then stops developing and simply waits for the next four months or so. Only around the time that the female dens up does the minute embryo implant itself on the uterine wall and resume its normal growth. This reproductive approach is known as delayed implantation, and it is found not only among bears but also among members of the mustelid, or weasel, family, including ermine, otters and fishers. It is thought that such a strategy allows the bears to mate during the time of year when travel is easiest and food abundant. Courtship and reproduction are, after all, strenuous, time-consuming activities, so better to deal with them at a time when they will cause the least disruption.

Pregnancy is another story. Bears have a gestation period of just ten weeks, so if the embryo implanted immediately, the sow would be giving birth in early fall—and her cubs would be facing winter while still barely mobile. By putting the entire process on hold for several months, bears and mustelids can enjoy the best of both worlds.

For an animal the size of a black bear, a ten-week gestation period is ludicrously short; by contrast, a hundred-pound whitetail doe has a pregnancy of more than six months. And while a deer will produce a fawn nearly one-twelfth her total weight, a female black bear gives birth to infants that are, in relation to their mother, among the smallest in the mammalian world.

Birth occurs in January or February, while the mother is in the slumbering throes of winter torpor. She rouses just enough to lick clean the babies, which look like nothing so much as small rats with short tails and a fuzz of black fur. Each one is about eight inches long and weighs about twelve ounces—about one-three hundredth of their mother's mass, for an average 250-pound sow.[5]

Black bears may have the strangest infancies of any Appalachian mammal. For their first three months they stay inside the den with their periodically slumbering mother, doing little but eating and growing. And they grow rapidly; by six weeks they have quadrupled their weight and have learned to walk. By the time the family leaves the den for the first time in late March or April, the cubs are fast, mobile and curious, weighing from five to eight pounds.

It is a beautiful and elegant system of rearing offspring. Because bear cubs are so tiny and helpless at birth, a female that bore her young in spring would be tied down at just the time of year when she needs the freedom to roam widely in search of food (a newly emerged bear will have lost as much as a quarter of its weight over winter). By giving birth at a time when she is stuck in one place anyway, the female can ensure that her cubs are old enough to travel with her from the moment she leaves the den in spring. And like the young of the great horned owls, the bear cubs have a tremendous head start. By the time they are a year old, they may weigh more than one hundred pounds, and by their second spring, they are ready to fend for themselves, while their mother enters another two-year reproductive cycle.

The other practitioners of delayed implantation, the weasel family, do not give birth in the dead of winter as does the black bear, but they push the limit of spring back as far as they can. Actually, in some respects they take delayed implantation as far as it can go. The river otter may mate within a day of giving birth, with a total gestation period of up to 380 days, although only during the final 60 or so are the embryos attached and growing. In a very real way, a female otter is pregnant almost all of her adult life.

Otters give birth in March or April, while fishers—another mustelid with a gestation period of more than a year—do so as early as February. Fishers, otters and other weasels, unlike bears, are active all winter, so the female faces the challenge of capturing enough food for herself and her rapidly developing fetuses. She must also find a secure den, one that will be safe even when she is on the hunt. For otters the choice is often an abandoned beaver lodge, a burrow in a riverbank or beneath a dry rockslide near water. For the fisher, which despite its nearly foxlike size is an agile climber, the choice is invariably in a tree cavity high above the ground.

The fisher is one of the most egregiously misnamed animals in the Appalachians, for it does not catch fish (although, ironically, it is attracted to the scent and has traditionally been caught in fish-baited traps). More than three feet long and weighing around twelve pounds, the fisher has a long, full tail and a catlike face, and looks something like a cross between a fox and a feline—hence the old mountain names "black cat" or "fisher-cat."

Once common through the Appalachians as far south as the Smokies, the fisher was among the cadre of wilderness animals done in by the loss of the virgin forest, although its demise was hastened in many areas by heavy demand for its lustrous, dark pelt, which fetched prices as high as three hundred dollars. Today it is found as far south as Pennsylvania and West Virginia. With protection and reintroduction efforts, the fisher has rebounded strongly and is again widely distributed in forested parts of the northern and central Appalachians.

Fishers are consummate hunters, large and strong enough to kill small deer on very rare occasions. Their normal diet, however, consists of birds and small or medium-sized mammals—snowshoe hares, red squirrels, voles, raccoons, grouse and the like, taken by ambush or at the end of a long, acrobatic chase. The species is most famous, however, for its taste in porcupines.

The question has always been: how? With more than thirty thousand quills, a porcupine is the most heavily armored animal in the Appalachian forests, and hardly what one would consider a subject for specialization.

Snow squalls leave the summit of Mount Abraham, in the Green Mountains of Vermont, covered in a fresh coat of white.

Yet the fisher, by its fondness for porkies, proves the rule that there is no foolproof defense in nature.

Folklore always credited the fisher with a lightning-fast sneak attack in which the porcupine was flipped over, exposing its unprotected belly. The truth is less imaginative. Only porcupines on the ground are bothered; those in a tree are safe. The fisher makes a direct, frontal assault, using its amazing agility to dart in at the porcupine's face, biting it around the nose and eyes, then backing away before the big rodent can spin around with its quill-studded tail. (The quills cannot, as many imagine, be thrown.) This lunge-and-feint attack goes on for some time, weakening the porcupine through injury and blood loss, until the fisher can safely close in for the kill. Only then does it upend the porcupine, opening the belly and rolling back the skin—quills inside—as it feeds.

Even the fastest fisher is going to pick up quills in its muzzle and forequarters during a fight with a porcupine. The structure of a porcupine quill makes even one a dangerous invader; the tip of the hollow quill is covered with minute, overlapping scales that point toward the base, and once the quill is embedded, muscular action tends to draw it deeper and deeper into the tissue. In some cases, quills have been known to migrate straight through the body, penetrating internal organs with fatal results. In most cases, however, a predator with a mouthful of quills dies of starvation before that stage.

Not so the fisher, however. Over tens of thousands of years of specialization, the fisher has developed a tremendous tolerance for porcupine quills, and most of the quills are diverted to the layer of subcutaneous fat beneath the skin. Even when the quills enter organs, either from outside or after being swallowed, the fisher seems to be able to withstand the intrusion. One fisher from British Columbia had five quills through its stomach wall and two through its intestine, yet appeared perfectly healthy.

One reason why the fisher is now protected and welcomed is its porcupine-eating habits. Across most of the North Woods porkies are considered a pest, damaging or killing trees as they feed on the inner bark. Fishers are seen as a natural control, and the results bear out that assumption. In parts of the Great Lakes region where fishers had been extirpated, porcupine populations dropped by more than 75 percent after the predators were reintroduced.

The woods around the beaver pond were completely still, save for the far-off lisp of a kinglet in the pines, a call so high and thin that it sounded like the squeak of snow underfoot on a subzero day. The Green Mountains were white silhouettes against the darkening clouds, slowly fading behind another drifting curtain of snow, and my breath smoked in front of my face every time I exhaled.

I walked out across the ice to the lodge, a shapeless mound beneath the new snow. The top was even with my neck, and I leaned over it, ear cocked, listening. There was no sound, of course. I've heard young beavers whimpering and muttering inside their lodges in summer, while I was canoeing, but now the layers of sticks and mud and ice and snow had sealed the lodge from the outside world. I could only imagine the beavers inside, huddled in warm, moist darkness, rubbing against each other, chattering their companionable, incessant chatter. They were insulated from winter, but I was not; my toes were getting cold, and the snow was getting heavier. I went home.

Wild Turkeys in White Oaks

Chapter 10

SEEING A FOREST FOR THE TREES

My Pennsylvania Dutch neighbors, whose ancestors settled this part of the Appalachian ridge-and-valley system nearly three hundred years ago, have an especially poetic way of describing our local mountains. The south slope, which gets the full brunt of the sun's rays, is the "summer side," while the cooler, damper north slope—regardless of the season being discussed—is the "winter side," the Germanic accent substituting a *v* for the *w* in winter.

The difference is more than semantic. Weeks after the spring thaw has melted the snow on the summer side, it still lies, dirty and sullen, on the winter slope. The summer side's warmer, drier ground supports hepatica and pink lady's-slipper orchids in the spring and sheep laurel in June; the winter side is where you look for squirrel corn and painted trillium, which like their roots cool, along with most of the species of ferns. Drought-tolerant mountain laurel grows abundantly on the south slope, moisture-loving great rhododendron on the north; the line of demarcation is easy to see in winter.

The most obvious differences, however, are in the trees. A damp area like a spring seep or the banks of a small stream will support a grove of tuliptrees on the summer side, while on the winter side such a setting is more likely to host a stand of hemlocks and white pines. The north slope's shade, lower average temperature and overall higher soil moisture allow a heavier mix of sweet and yellow birch, red maple, striped maple and other cool-climate hardwoods; this is especially obvious in autumn, when their vibrant foliage makes the winter side the more colorful of the two.

Summer side or winter, however, oaks dominate these Pennsylvania mountains. There are about a half-dozen species that are found commonly in the hills around my home, and they are divided into two great clans: the red oaks, whose leaves sport sharply pointed lobes tipped with bristles, and whose acorns are laced with tannin; and the white oaks, whose leaves ripple with softer contours, and whose acorns are sweeter, without as heavy an acidic load. Red oak acorns require two years to mature, white oaks' only one. Red oaks have generally dark bark, white oaks generally light. They are polarities within one genus.

Among the red clan, the northern red oak is the most common species hereabouts, a tall, robust tree that grows quickly, its upper trunk patterned with smooth, vertical strips that shine in the sun. The leaves are sharp and angular, with between seven and eleven pointy lobes; the scarlet oak, which grows less commonly around here with the red, has similar leaves, but the lobes are somewhat narrower and the sinuses—as the gaps between the lobes are known—are deeper. Scarlet oak is not quite so tall a tree; while a mature red oak will average seventy-five or eighty feet, a scarlet is lucky to hit fifty or sixty.

Red and scarlet oaks can tolerate warm, dry conditions, but on the highest, driest ridges the chestnut oaks come into their own. In kinder conditions these oaks, named for their long, gently lobed leaves that resemble a fat chestnut's, may grow a hundred feet tall, but few on the ridgetops make it to such heights. These are tough, gnarled trees that scrabble for a living in meager soil among the boulder outcroppings, rarely growing very tall, rarely growing very straight. The bark is thick and deeply furrowed, and it is so rich in tannic acid that in the nineteenth century "rock oak," as it was called, was cut and stripped naked, its bark sent to the tanneries for use in curing leather.

None of these oaks can compare with the eastern white oak, however. In almost every respect—size, age, nut quality, even the subjectivity of aesthetics—white oak is a standout. The name comes from the pale, creamy wood, but it fits from the outside, too; the trunk is covered with a fine, scaly bark of pale gray, the color of a deer's midwinter coat, or a slab of sandstone, or a foggy dawn. Young white oaks—and in this long-lived species that can mean trees already a century old—stand straight in the forest, thickening with the years like Doric columns.

In the forest, an oak is forced to grow tall and straight in the battle for sunlight, but a white oak planted in the open (or spared when all around it are cut) develops a superbly rounded crown, with a short, very thick trunk and great, low branches reaching straight out. No other tree gives such a visible sense of its inner strength, such a look of muscular power. Some pampered,

Two pink lady's-slipper orchids bloom against a backdrop of white pine seedlings.

ancient specimens, usually on old estates along the coastal plain, may have a girth of more than twenty feet and a crown spread of forty yards or more.

In the valleys between the ridges, big white oaks are common only along the edges of fields, and in years past, they were permitted to grow out in the fields themselves, a welcomed spot of shade during the heat of summer mowing. But with larger and more sophisticated farm machinery, such trees are suffered less and less often, for they break the regularity of the plowing. The map of memory that I carry with me is marked by these long-armed giants, and I feel a stab of genuine grief whenever I crest a hill and see the rounded, empty swell of a field punctuated by a new stump, where a century-old white oak once grew. It's as though a piece of my past has been cut up for firewood.

Many of the venerable white oaks are no longer visible from a distance. There is less farmland and more woods around here than there was a hundred years ago, and a good many of the old field trees have been absorbed into the forest, where they stand like grandparents among the younger generations, short and broad against the screen of straight, light-hungry offspring. Old and large as they may be, such oaks are striplings, rarely more than a century and a half old. White oaks are among the most long-lived of eastern trees, and ages in excess of six hundred years are not unknown. Placing a hand on the corrugated skin of such a Methuselah, it is impossible not to sense the swirl of history that has passed beneath its limbs.

About twenty miles from my home, along the headwaters of a small trout stream, the oldest white oak in the area once grew. Most estimates put its age at 250 or 300 years, although its trunk had been hollowed by decay, so there was no way to take a boring and make a ring count, and any age was just a guess. In truth, the tree was a disappointment to the untrained eye, for time and bad weather had taken all but one of its huge branches, and it was only barely alive. A screen of tall rhododendrons blocked it from view until you were right on top of it, and it had a lopsided, damaged look—not the heavy-limbed symmetry of youth, but a sadly withered elder in its dotage.

The few times I took others to see it, I could trace the letdown in their eyes. They were expecting a monarch and instead found wreckage. But if you could look beyond the moment to the past, it was a remarkable tree. In its presence I was free to imagine the world as it used to be, when this was the hunting ground of the Lenape, and the air thundered with flocks of wild pigeons.

The tree fell a few years ago, brought down by a thunderstorm, a perfectly natural death that old trees are rarely granted these days. Lying on its side, the empty cavity of its heart was exposed to the sun, with just a thin rim of solid wood around the outside; I thought it a wonder it had lasted as long as it had. Such is the tenacity of an oak.

Closely related as they may be, the various species of oaks all have their own set of environmental requirements, which in concert determine just where they grow. Soil fertility, drainage and climate are obvious factors, but one that humans frequently overlook is light.

Sunlight would seem to be the most democratically distributed of all a plant's needs. Rain may come and go in flood or drought, soil may be miserly or rich, but the sun touches every part of the Earth with exactly the same amount of light every year; the short days of northern winters are compensated for by the shorter nights of summer.

Yet the battle for light is one of the dominant forces in the existence of a forest plant. Each species has a minimum (and in some cases a maximum) level of sunlight needed for survival; beyond that threshold it will not thrive, and may even die, for it will not have enough energy for photosynthesis. And the lower a plant grows in the forest, the harder it is to get enough light.

A mature, well-developed deciduous forest will have five distinct layers: the highest canopy of treetops; a subcanopy made up of younger trees of the same species, waiting for a gap to open; an understory of smaller trees like dogwood, adapted for growing in the shade; a shrub layer of plants like wild azalea and blueberry; and a ground layer of herbaceous plants, ferns, mosses and lichens. As any photographer knows who's tried to work in the gloom of a robust hardwood forest in summer, the layers are expert at catching most of the available light. At ground level, the light intensity may be less than one–four hundredth of that falling on the canopy.

Plants that live far down in the forest layers must make do with this paucity of solar energy, or find ways around it. Many wildflowers, known as spring ephemerals, take advantage of the fact that the trees need more than a month of warm weather to fully leaf out. These flowers, which include Virginia bluebells and trout lily, make a habit of sprouting with the first hint of thaw and finishing their entire reproductive sequence before the tree leaves appear; by June the bluebell and trout lily leaves have already yellowed and withered away.

Other vernal wildflowers keep this same early schedule but retain their foliage through the summer, able to withstand the lower light levels once blooming is out of the way; the several species of trilliums, wild bleeding heart and squirrel corn are among this group. Still others, like wood sorrel, Solomon's-seal, baneberry and many ferns—known as shade-obligates—cannot take the full sun of the naked spring canopy, and wait until the tree leaves arrive before they open their own leaves.

A fern may be able to eke out an existence in the leftovers of light that reach the forest floor, but oaks need the full lash of summer sun to reach their greatest potential. For a ninety-foot red oak this is no problem, but the obstacles are much greater for its seedlings.

Masses of large-flowered trillium blanket a Pennsylvania hillside, taking advantage of the bright spring sunshine before the tree canopy closes, plunging the forest floor into constant shade.

Four of the common Pennsylvania oaks—scarlet, red, chestnut and white—exhibit a gradient of tolerance for shade that affects where and for how long they can grow. Scarlet oak is the least shade-tolerant of the bunch; it does well in open woods or as a pioneer species in a recently cleared area, where its rapid growth gives it a competitive advantage. Once the canopy closes, though, its seedlings are restricted to those small gaps where sunlight makes it to the forest floor. Consequently, scarlet oak tends to become uncommon in older, more mature forests.

Northern red oak is a bit more forgiving of shade than the scarlet oak, and so retains its position as the forest moves toward the relatively stable stage of succession known as climax. Chestnut oak, too, can withstand shade competition—the forestry term is overtopping—from other trees. White oak is the most shade-tolerant of all, sprouting fairly well even under canopy, and can thus maintain or even solidify its position in the forest as time goes on. The saplings bide their time, waiting for an opening, then race toward it, plugging the gap before another tree can exploit it.

The seedlings also have a few adaptations to help them through the lean years. You can easily see this for yourself by comparing the size and shape of the leaves from a sapling oak with those from a mature specimen growing in the open. Almost always, the sapling's leaves are larger, with less pronounced indentations—both characteristics that create more surface area with which to capture sunlight. Most wildflowers of the deep-woods floor, like trilliums or jack-in-the-pulpit, exhibit exceptionally broad leaves for the same reason.

In the ridge-and-valley system of the Pennsylvania Appalachians, especially on the southern "summer side" of the hills, oaks account for as much as three-quarters of the trees in the woods. This is an unusually, even unnaturally, high percentage and stems from two factors. When the mountains were cleared during the eighteenth and nineteenth centuries, the original mix of diverse hardwoods and abundant conifers was replaced by a more solidly oak forest, since oaks sprout quickly from stumps (conifers do not) and have seedlings that can capitalize on sunny, disturbed land. Also, the chestnut blight robbed this region of its other dominant hardwood, leaving oaks to fill the space.

Any time you have an ecosystem composed largely of one species or a group of related species, you're inviting trouble; a single disease or pest can sweep through uninhibited, whereas in a more diverse community, its effects would be minimized. That's why farm crops require such elaborate coddling. The central Appalachian ridges are so heavily populated by oaks, in fact, that they can be seen as something of a monoculture, much as an agricultural crop. Anything that affects the oaks will necessarily have a great impact on the entire forest system.

Enter the gypsy moth, a native of Eurasia with a passion for oaks. It was brought to Medford, Massachusetts, just outside Boston, in the mid-nineteenth century by French entomologist Etienne Leopold Trouvelot, who was trying to produce hybrid silkworms and establish a silk industry in the United States. (The choice of gypsy moths seems strange, even for hybridization, for this species spins no silk cocoon when it pupates.) In 1869 several of Trouvelot's gypsy moths escaped into a neighbor's woodlot; he is said to have pleaded with the owner to burn the trees, but the man refused.

If ever a fire would have been a good idea, it would have been then. The gypsy moths had been set loose on a continent of oaks, and all of their natural enemies, diseases and parasites—the normal checks and balances for any organism—had been left on the other side of the Atlantic. To say the gypsy moths exploded is a poor understatement; within forty years they had overspread New England and New York, defoliating huge areas each summer, and were turning south, into the oak-rich heartland of the Appalachians. Through the last century they continued to ripple outward, south and west. The tidal wave rolled over eastern Pennsylvania, where I live, in the late 1960s and early 1970s.

The name aside, it is not the moth stage that causes problems, but the caterpillars. They have an insatiable appetite for leaves—more than three hundred species of trees and shrubs, but with a particular fondness for oaks. The fuzzy, cream-colored egg masses, glued to the sides of the trees, begin to hatch in early May just as the leaves are unfurling, and the minute caterpillars, each barely an eighth of an inch long, spin long strands of silk. Caught in the wind, they may "balloon" for miles, much as young spiders do, before settling down to eat. For this reason, the windward sides of ridges are often the hardest hit.

The caterpillars grow quickly, periodically molting their skins along the way. Eventually they reach a length of more than three inches—each one a blue-gray larva with large, speckled-yellow eyes, dots of red and blue down the back, and a protective screen of long, irritating hairs clumped along the body. These hairs are an effective defense, and with the exception of cuckoos, few birds will touch a large gypsy moth caterpillar.

During an infestation there are so many gypsy moth caterpillars in the trees that you can actually hear the subtle rustling of their chewing, with a background patter of falling dung pellets (known as frass) that sounds like a gentle rain. The trees are stripped bare, and the ground is covered with a green snow of leaf fragments that the caterpillars missed; as the food runs

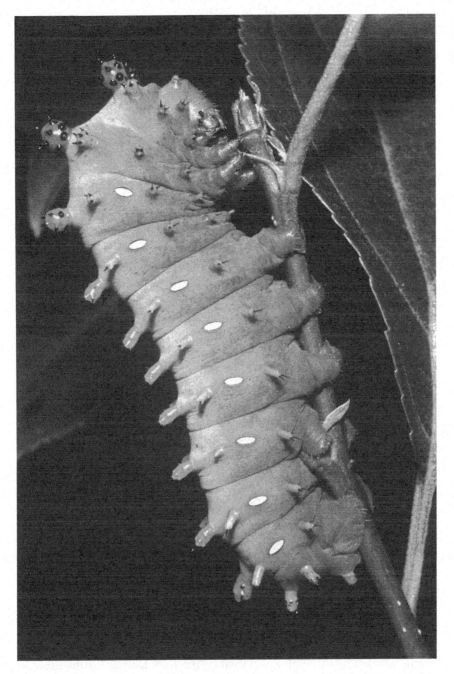

Plump, green and big as a man's thumb, a cecropia moth caterpillar feeds relentlessly, preparing for its transition to the largest moth in the Appalachians.

out in the canopy, many creep down to the ground to consume even these leavings.

I have lived in the midst of several gypsy moth explosions, and it is a disturbing, uncomfortable experience. For several weeks the caterpillars are everywhere—covering the trees, flowing up the sides of buildings, marching across roads in columns so thick that their smashed bodies are as slick and dangerous as ice. The humid summer air has the acrid tang of ammonia from their decaying droppings, and the temperature at ground level soars. Wood thrushes that built their nests in what, just two weeks before, had been shady, sheltered spots now find themselves shielding their chicks from the June sun, panting in the heat. Some people break out in rashes and hives, the result of the tiny hairs that break off from the caterpillars' bodies and drift on the wind. If a larva drops from the trees and hits the back of your neck—an almost inevitable experience for a hiker—the hairs leave an itchy, gypsy moth–sized welt where they pricked the skin.

By mid-June, the caterpillars reach full size and pupate. They move to sheltered locations—beneath picnic tables, inside hollow trees, between the corrugations of the bark—and molt into a dark brown pupal case, dangling from its tail end. Over the next couple of weeks the moth's body undergoes radical alteration, so that when the shuck splits, an adult moth emerges. Males are brown and capable of flight, but the creamy females are too glutted with unfertilized eggs for travel. Instead they broadcast a powerful pheromone called gyptol on the wind, luring males to mate. Once the furry egg masses are laid, both sexes die.

Gypsy moths are cyclic, moving from boom to bust on a nine- to eleven-year sequence. At the peak of the infestation the ridges are winter-brown, and the starving caterpillars may even try to eat green paint from the sides of buildings. Such excess cannot last, and the end comes quickly. Starvation opens the door to a viral disease called wilt, which kills the caterpillars by the hundreds of millions, their bodies jackknifed on the tree trunks, clinging tenaciously even in death. Now the forest has the stink of decay for a week or so, but there are almost no adult moths to follow. For several years the sight of a gypsy moth is rare, until the population begins to build again.

And what of the trees? Looking out over a denuded area, the damage appears catastrophic, but it isn't quite as bad as it looks. Unless an oak is already stressed by drought, disease or competition, a single defoliation will not usually kill it, although it will weaken the tree, leaving it vulnerable to attacks by other pathogens (back-to-back defoliations may be fatal, however). Within a week or so after the caterpillars pupate or die, the oaks begin to put out new leaves—not so large as their first set, and never so many, but the flush of fresh green comes as a great relief to the forest.

Some oaks will die, however. If they've been stripped two years in a row, the losses may be as high as 30 percent, even more if the attack coincides with a drought, or infestations by other pests like sawflies or forest tent caterpillars. For a single defoliation in a year of normal rainfall, however, the usual death rate among oaks is about 5 percent. Other trees like conifers, which may be eaten in desperation by starving caterpillars, may succumb after just one attack.

When gypsy moths first appeared in the central Appalachians, many people predicted the destruction of the forests. I recall seeing the empty, denuded trees and the brown mountains for the first time and agreeing with that gloomy assessment. This was just at the time when national opinion was coming down against such "hard" pesticides as DDT, but the arsenal of sprays used against the gypsy moth was still formidable—persistent chemicals like Sevin and Dylox that killed a wide spectrum of insect life, not just caterpillars. Huge areas of Pennsylvania were sprayed, sometimes over the objections of landowners. Hawk Mountain Sanctuary's two thousand acres of ridgetop were sprayed in 1968, despite pleas from the sanctuary; it was years before a honeybee was seen in the mountain meadows again, and bee populations never really recovered, although the gypsy moths bounced back quickly.

In fact, chemicals could not eradicate the moths, and in some ways their use may have prolonged the agony. Spraying kept the caterpillar population below the critical level needed to trigger starvation and wilt epidemics, and also killed the natural predators and parasites that had been introduced to this country as biological controls. This may have exacerbated the stress on trees by stretching near-defoliation periods out for more than one year, while playing hob with the underpinnings of the forest food chain.

Slowly, bureaucrats came to recognize that the gypsy moth was here to stay, and that statewide spraying was an expensive, environmentally dangerous boondoggle. The state switched from chemical sprays to a biological agent, a bacteria known as *Bacillus thuringiensis* (Bt for short), which it now applies only to wooded residential areas, state parks and other high-use areas. While an improvement, Bt is not perfect. Sometimes billed as a specific killer of gypsy moths, it actually kills any moth or butterfly caterpillar that ingests it.

Ecologists, meanwhile, see the complexities of the gypsy moth. No one is glad it's here, and there is no doubt that it kills marketable timber, reduces cover during the peak of the nesting and breeding season, weakens trees—a whole litany of ills. But remember that the gypsy moth is a symptom as much as a disease—a symptom of the unnaturally heavy concentration of oaks in the forests of the central Appalachians. Although the caterpillars will eat dozens of species of trees when pressed by hunger, they can only reach plague proportions if they have an abundance of oaks; many native hardwoods are unpalatable to them, and some, like white ash, tuliptree, red maple, dogwood, cherry, bitternut hickory and spicebush, are left strictly alone even by starving caterpillars.[1]

This boom-and-bust cycle was finally short-circuited by a fungi first discovered in the bodies of dead gypsy moths in Japan and brought to Boston in 1910 amid high hopes—whereupon it vanished without a hiccup, and was declared a failure. In the mid-1980s scientists again tried releasing spores of this same caterpillar-killer, *Entomophaga maimaiga*, and again it seemed to do nothing—until gypsy moth larvae in Connecticut started dying three years later. Ironically, DNA tests suggest that *Entomophaga* finally established itself by accident, not through the deliberate introductions, a rare example of a happy outcome from an invasive alien. However it arrived, the fungi has proven to be an extremely effective biocontrol, spreading rapidly throughout most of the gypsy moth's ever-expanding range and damping down the massive defoliation events that were once routine, Best of all—so far at least—it seems to affect only gypsy moths.

It was nearly midnight on a frosty, full-moon February night, when the old patches of snow on the frozen ground seemed to shimmer with their own light, and the leafless oaks stood out in stark silhouette against them. I was returning from a late visit with friends, and my route led me over the Kittatinny Ridge. I was tired and getting a little drowsy, so on impulse I stopped the car near the top of the mountain, pulled on a hat and gloves and took a walk through the woods to clear my head.

The short trail led about two hundred yards to the edge of a rocky drop-off, with the toe of the valley coming to an end down below—a bowl of moonlight and cold cradled between the hills. I listened for great horned owls, but it was egg-laying time, and the owls become quiet and secretive at that stage. A screech-owl whinnied downslope a long way, and I tried to whistle back, but the cold had numbed my lips so that my imitation only silenced the bird.

The moon had washed the night sky clear of most of the stars, although I could dimly pick out the Great Bear to the north. In the silence, as I looked at the stars, I heard a weird wail rise from the woods below me.

It's hard to describe the sound that followed. It was a single voice, breaking almost immediately into a falsetto that wavered and dropped off, then picked up the volume again in a loud, piercing note that rang momentarily before ending with a chopped abruptness. The interlude lasted a few heartbeats, then the banshee began again, howling up and down the scale. Even though I knew what it was, the hair on the nape of my neck bristled with gooseflesh.

The coyote lapsed into silence again, and I cupped my hands to my mouth, tipped back my head and howled as best I could. Before I'd even finished the coyote answered, its voice climbing an octave above mine in an unearthly harmony that made me shiver again to hear it. When I paused for breath, the howls still hung in the black, cold air like a summons, insistent and wild.

Sound is the soul of the night—the snort of a deer, the yowling scream of a bobcat or the caterwauling of a gray fox. Wild sound defines the existence of unseen animals in a world of moonlight and shadows, stamping the spot as a haunt. A twisted pine is just a twisted pine, until a great horned owl thunders a string of bass notes beneath Orion's eye. Then it becomes an owl tree, forever marked so in the map of the mind. A wooded valley is just a valley, until a coyote sings its night song to the moon; thereafter the place is special.

The music of coyotes is lately come to the Appalachians. Only within the past seventy years have these mountains, which have lost so much, gained this newest voice. There are no clear records from the eighteenth and early nineteenth centuries of coyotes in the East; some ambiguous reports of "brush wolves" may refer to coyotes, but the term was also used for canids that were by their actions probably eastern wolves or red wolves (in the southern Appalachians, such animals were almost certainly red wolves). It is assumed that if coyotes were present, they were kept at extremely low levels by competition with wolves.

Wolves were eliminated in many parts of the Appalachians rather quickly, however, and coyotes did not appear until the middle of the twentieth century, reinforcing the belief that they are true immigrants, expanding from the west. Most scientists believe they looped up around the Great Lakes through Ontario, then southeast into the United States again. They were first noticed in New York in the 1920s, and in northern New England a decade later.[2] Today they are found in virtually every part of the East, and in the entire Appalachian system from the Gaspé south.

There had been rare reports of coyotes in Pennsylvania since the first years of the last century. It's unclear where they came from, although some may have had human help. During World War II, when the huge military installation at Fort Indiantown Gap was bursting at the seams, western GIs supposedly brought coyote pups East with them, then released them to the hills when it was time to ship out. If any survived, it was at a very low population level, and perhaps they hybridized with domestic dogs. Sometime in the 1970s, however, reports of coyotes cropped up with increasing frequency all over the state, especially in the heavily wooded, lightly populated mountains of the northern plateau region and the Poconos.

This was a different animal from the western coyote. For one thing, it was considerably bigger. A western coyote averages about twenty pounds, while a male eastern coyote will run to thirty-five or forty, and a few have been recorded at more than sixty pounds. At one time this was thought to be a result of crossbreeding with domestic dogs—so-called coydogs—but recent genetic studies have confirmed that eastern coyotes carry no dog genes, but are a mix of coyote and eastern (timber) wolf parentage, a relic of the western coyote's expansion north and east through southern Canada, entering what was already a canine mishmash of eastern wolves, gray wolves and their hybrids.

What to call the unique animal that emerged, spreading east and south across the Appalachians, remains an unsettled question. "Eastern coyote," the most common label, ignores the creature's wolf heritage, but names like "tweed wolf" (used in Canada), while more poetic, are equally inadequate. "Eastern coywolf," the term suggested by the team conducting the genetic work, hardly rolls off the tongue, but it's better than "eastern canid," a cinderblock of a name that had been in limited use. Whatever you call this big and adaptable hunter, in a small but very real way the Appalachians have regained their wolves.

The first eastern coyote I saw was dead, shot by a neighboring dairy farmer on the opening day of buck season. It was a male that weighed forty-two pounds, with an amazingly thick, lustrous coat, and was hanging by a hind foot in the man's barn. In the harsh light of a bare electric bulb, inside the whitewashed butcher shop, it looked, if anything, bigger than life. The fellow shot it because he feared for his cattle, and there are occasional reports of livestock attacks, mostly on sheep. By and large, however, coyotes have been quiet neighbors. Studies by academic researchers and the Pennsylvania Game Commission show that they feed most heavily on mammals (woodchucks, mice and rabbits top the list), ground birds, scavenged carrion, berries and crops including apples and pumpkins. Some deer are killed, although more often they feed on roadkills and the gut piles left by hunters. They have also shown a fondness for domestic pets, although the dogs may be killed more out of territorial rivalry than hunger.

Coyotes are now common essentially everywhere in the East, and have grown comfortable in urban areas, farms, and forests. I see their tracks and scat on a regular basis, although I rarely see the animals themselves. When I do, it is usually a fleeting glimpse across the headlights or a brown shape slipping quietly away through the laurels.

Wherever they live, coyotes have always been a lightning rod for emotion, much of it negative. In northern New England, particularly Maine, anti-coyote sentiment is fueled by concern for the deer herd. The same thing is occurring in rural areas of Pennsylvania, especially in the northern counties where the whitetail herd has decreased in recent years—not from coyote predation, but as part of a calculated increase in hunting pressure by the state, designed to bring the deer herd into line with its habitat. Regardless, calls for bounties have grown (thankfully falling on deaf ears), and some sportsmen's clubs have taken to organizing midwinter coyote hunts, with prizes for the biggest females. But coyotes have been evading humans for years, and the results are usually lopsided; one recent hunt drew 217 sportsmen, lasted two days and yielded just one coyote.

A couple of years ago, on the opening day of deer season, I was sitting quietly along the top of a ridge watching several trails below me, hoping to intercept a buck. I saw a tiny movement to the side and felt my pulse surge, but it was no deer. A coyote appeared among the laurel and blueberry bushes, moving at a steady trot; the wind was in my favor, and the animal passed within thirty or forty feet of me without ever noticing my presence.

The coyote was fairly small—a female, most likely—and rather scruffy looking. She had a dark triangle of blackish guard hairs on her back like a cape, and her tail was full and also black-tipped, hanging behind her at half-staff; the front of each leg had a dark stripe, a trademark of the eastern race. She never slowed, never even looked my way, and I moved only my eyes as she passed, so I wouldn't give myself away. That night I talked to two friends, both deer hunters like me, and told them about my encounter. One man asked bluntly why I hadn't shot the coyote. The other just said "Wow" in a quiet, envious way.

On several occasions, traveling in remote areas of the upper Amazon, I've seen birds that didn't quite fit anything in the field guides. This may be because there were at the time few decent bird guides available for that region, or (even more likely) that I just had a bad view. But I'm always struck by the possibility that I might have been looking at an unknown species.

Such a feeling is rare today, even in the most far-flung corners of the world. In the Appalachians, where the bird life has been plotted for the past three centuries with growing precision, there would seem to be very little mystery left. Yet there are a few puzzles still to solve. One that I have always found especially intriguing is the question of Sutton's warbler.

There are thirty-three species of wood warblers from the family Parulidae that breed within the Appalachians proper; this is the single largest and most diverse family of birds in the region, just nudging out the assemblage of finches and sparrows known as the emberizids. Most eastern warblers were described to science—the official "coming out" process by which a new species is formally acknowledged—in the early 1800s. Those were the days of Audubon, Alexander Wilson and other pioneering ornithologists, who roamed the woods collecting (that is, shooting) specimens and slapping names on those that struck them as previously unknown. While many were genuinely new, some of them, it turned out, were the females or immatures of previously recognized varieties. Those are the unavoidable pitfalls of describing the fauna of an unknown country—there are no handy reference works to fall back on.

In any case, the dust settled rather quickly. By the latter half of the nineteenth century the mistakes had largely been sorted out, and the search for new birds had moved west. While there were still many questions about the distribution and dynamics of Appalachian bird life, there were few unknown birds left.

Then, in 1939, came the electrifying news that a previously undescribed warbler had been found in West Virginia. On May 30 of that year, Karl W. Haller and J. Lloyd Poland had been searching for birds along Opequon Creek, in the island-like panhandle near Martinsburg. Stopping to listen to a winter wren, they noticed a peculiar song—like a parula warbler's ascending, buzzing trill, but repeated twice rather quickly. When they located the bird, Haller and Poland at first thought it was a yellow-throated warbler, although that species had never been seen in the area, but then realized that it didn't precisely match any known species. They shot the bird for a scientific specimen. Two days later, about eighteen miles away in a sycamore forest along the Potomac River, they collected a female of the same type; worn feathers suggested she had been nesting.

Haller and Poland named their find Sutton's warbler for George Miksch Sutton, a noted ornithologist and Haller's teacher, and gave it the scientific name *Dendroica potomac*. The trouble was that not everyone was convinced they had discovered a new species. The male, skeptics noted, had a mix of features from two recognized species—the yellow throat, black-and-white facial pattern and white tail spots of the yellow-throated warbler, and the greenish back patch and basic song characteristics of a parula. More likely, it seemed, that the mystery bird was a hybrid between these two common species.

Parulas were relatively common in eastern West Virginia at that time, but yellow-throated warblers were not—in fact, it was not for decades after the discovery of Sutton's warbler that anyone found yellow-throateds in the area at all, bolstering the hybrid theory. For years, no other specimens came to light, which only reinforces the staggering odds that Haller and Poland bucked; hybrid or not, collecting two unknown birds within two days of each other in widely separated places is profoundly unlikely. There have been sightings in the decades since, mostly in West Virginia but a few farther afield, and in 2008 a female (showing signs of breeding) was captured and banded in Pennsylvania. All the records came within the wide zone in which parula and yellow-throated warblers overlap.

Even those who believed *Dendroica potomac* to be a hybrid acknowledged the possibility, however slim, that it might be a genuine species—perhaps naturally rare and within a hair's breadth of extinction when it was discovered. Others, including eminent Appalachian naturalist Maurice Brooks (who later saw a Sutton's warbler along Opequon Creek in 1942) thought it might be a newly emerging species. More recent evidence, however, strengthens the case for hybridization, including the discovery of a presumed second-generation bird that resembled a yellow-throated warbler but sang a Sutton's song.

Sutton's warbler aside, there may well have been Appalachian birds that winked out just as science was taking note of them. Students of American bird life have always been puzzled, for instance, by several of John James Audubon's paintings. A vain, self-aggrandizing man, Audubon was nevertheless an acute observer who added immeasurably to the early understanding of North America's avian fauna. That's why it is hard to completely dismiss his trio of "mystery birds"—the carbonated warbler, the Blue Mountain warbler and the small-headed flycatcher (despite the name, also thought to be a warbler).

Audubon shot two males of the carbonated warbler in western Kentucky in 1811; the yellowish, dark-capped birds shown in his painting are now believed to have been first-year male Cape May warblers. There's no consensus on the Blue Mountain warbler and the small-headed flycatcher, which are portrayed together in another painting. The warbler is, to my eye, most like a young female cerulean warbler—yellowish underparts, greenish above, with two white wing bars, although it has tail spots, which the cerulean lacks. The "flycatcher" is even less flamboyant, a drab, olivish bird with thin wing bars and a nondescriptive air to it.

All three species were based on collected specimens, but over the years the stuffed skins vanished, and with them any hope of solving the puzzle. The same is not true of the only specimen of Townsend's bunting, a sparrow-like bird collected only once, in southeastern Pennsylvania in the 1830s. A slice of history, its skin rests in the Smithsonian, although the puzzle remains. Some thought it a real species, others a hybrid between a dickcissel and a blue grosbeak; the former was once a common species in the East, but the latter was at best very rare in the state. The latest consensus is that it may have been a dickcissel that lacked the normal yellow pigmentation on the head and breast—a conclusion bolstered by the discovery, in Ontario in 2014, of what appeared to be the first "Townsend's bunting" in almost two hundred years. Though photographed, that bird vanished before a feather or blood sample could be taken to finally solve this two-century-old mystery.

Unlike the mystery warblers, which display a mix of plumages from several species and could, therefore, be hybrids, the bunting looks unique enough that it might have been a true species. And why not? At the start of the nineteenth century, ornithology was taking its first, halting steps, but habitat destruction had been proceeding merrily for close to two centuries east of the Appalachians. The damage was even greater in the West Indies, where several species of warblers winter. Who knows what rare, local species of animals and plants might have vanished before anyone with scientific training had a chance to notice them?

In 1870, an amateur ornithologist named Harold Herrick shot an unknown warbler in New Jersey, a yellow bird with black on its face and throat. He named it Lawrence's warbler in honor of a friend. That same year, Massachusetts ornithologist William Brewster collected another unknown warbler, a blue-gray bird with a yellow cap and yellow breast band; it was named for Brewster himself when it was formally described four years later.

Lawrence's and Brewster's warblers were not, in fact, true species. They were (and are) hybrids, as ornithologists came to realize, between two common types of warblers, the golden-winged and the blue-winged. They would be little more than a scientific footnote, except that the process that creates them continues in the Appalachians, providing avian biologists with an unmatched opportunity to see what happens when natural competition runs its most extreme course.

Around the time I graduated from high school in the 1970s, my parents bought a few acres of old farmland that were bordered at one end by a long-neglected orchard. Among the apples had grown up a thicket of bush honeysuckle and goldenrod, and in spring, I could sit by the edge of the weedy orchard and hear, drifting through the air, a chorus of *bee-bzz-bzz-bzz*, the songs of golden-winged warblers.

The males were natty little dazzlers, blue-gray with black masks and throats offset by white, with deep yellow on the top of the head and the shoulders. The females were a washed-out version of their mates, as though they'd started out the same but had bleached in the sun. They were the jewels of the old orchard, and I loved seeing them. I use the past tense because the goldenwings are gone from that overgrown orchard. Within the space of only a few years, they dwindled and disappeared—their song replaced by that of their near-relative, the blue-winged warbler.

Blue-winged warblers are members of the same genus as the goldenwings, and in size and shape they are almost identical; even their song is similar, a *bee-bzz* instead of the goldenwing's *bee-bzz-bzz-bzz*. DNA analysis suggests they diverged into separate species some 1.5 million years ago, and in plumage they are strikingly different. Both bluewing sexes are yellow, with a thin black line through the eye, blue-gray wings with white wing bars and a greenish back.

As with salamanders in the Smokies and darters in Appalachian rivers, what drove these two close relatives in such different evolutionary directions was probably isolation. Originally, blue-winged warblers were uncommon residents of the south-central states west of the Appalachians, while goldenwings occupied the higher elevations of the mountains south to Georgia, through the Northeast and upper Great Lakes region. Both preferred old fields and brush lands, but in the presettlement era their ranges were separated by wide areas of forest.

With agriculture's arrival the forests fell, replaced by cultivated fields, but these were scarcely more hospitable to either species than the original woodlands, and both remained rare or uncommon in many areas. But with the passing of the first farming boom in the 1800s, many fields were allowed to revert to trees, and the sudden proliferation of weedy habitat was a gift to both warbler species.

The solid forest that had once separated bluewings from goldenwings, however, was gone. Blue-winged warblers began a rapid expansion to the north and east, encroaching into what had always been goldenwing territory. Ordinarily an invasion by one species would have little impact on other birds, but while isolation had molded the bluewings and goldenwings into species that looked and sounded different, it had not changed their ecological requirements.

Both birds live in much the same kind of habitat, the tangled, early successional period between open field and young forest. They eat much the same foods captured in much the same ways—arthropods, including a large proportion of spiders, found while foraging in clusters of dead leaves. Both species nest on the ground in cups of woven grass, although here bluewings show a preference for slightly moister situations, and—this may be critical—a tolerance for somewhat older, taller brush than the goldenwings. In other words, goldenwings need fields that have been lately abandoned, while bluewings can take advantage of a wider range of habitats.

All in all, though, bluewings and goldenwings are two species trying to occupy roughly the same niche. They are in direct competition with each other, and something must give. What gives are the goldenwings.

Ever since the blue-winged warblers began expanding north, the goldenwings have been retreating. Bluewings are the more aggressive of the two—not more pugnacious or quarrelsome, necessarily, just more adept at occupying their preferred niche, so that they quickly displace the goldenwings already there. Biologists have plotted this rate of incursion and find that, on average, bluewings completely replace goldenwings within fifty years of first contact—an eyeblink from a geologic perspective. In small, local settings like the old orchard at my folks' place, the change can come even faster.

Because they are so closely related the two species easily hybridize, producing fertile offspring. Bluewings, as it turns out, are dominant genetically, too. The first-generation hybrid shows a predominantly bluewing pattern, with a thin black eyeline and plain throat, along with yellow on the top of the head and chest—the "Brewster's warbler." In the early years of contact, this form is the most common. Only later, as Brewster's hybrids backbreed into the increasingly swamped gene pool, do the rare "Lawrence's warblers" begin to appear, their goldenwing pattern suffused with bluewing yellow. This is the result of recessive goldenwing genes bubbling to the surface, even as the pure goldenwings disappear from the scene altogether.

I've seen a number of Brewster's hybrids over the years, including one that had mated with a female bluewing and filled a nest with genetically suspect eggs. Lawrence's hybrids are much rarer, and I've never been lucky enough to see one. I may never at all, because the blue-winged wave has rolled through my area already, and goldenwings are few and far between.

Their roots deep in soggy soil, false hellebore lines a slow-flowing creek in Cranberry Glades, West Virginia.

No one's sure where all this will lead. The golden-winged warbler has experienced one of the steepest population declines of any North American passerine, and for a time, it appeared that its extinction might be imminent. But it has also undergone a dramatic geographic shift—vanishing from much of its southern and central range, while expanding dramatically to the northwest, into places like Minnesota, Wisconsin and Manitoba where the majority of the species now breeds. It also seems to do better at higher elevations than the blue-winged warbler, and responds remarkably well to experimental timber management. Perhaps the goldenwings will strike a balance with the interlopers. All I know for sure is that I miss the reedy *bee-bzz-bzz-bzz* that used to float through my spring dawns.

There is growing concern among conservationists not just for the golden-winged warbler, but for a whole spectrum of Appalachian songbirds, including some of the best-loved—thrushes, orioles, vireos and warblers. Birders had been saying for years that the spring and autumn migrations just weren't what they used to be. The eagerly awaited "warbler waves" of spring, when treetops would fill with thousands of weary migrants finishing a night of travel, became fewer and farther between, and the number of birds in the waves seemed to drop year after year. At first, these complaints were dismissed as nostalgia for the good old days. Casual observations are dicey barometers, after all, since they are subject to so many variables. Only in recent decades have scientists been able to say with any certainty that the suspicions are right—the decline is real, and of potentially disastrous proportions.

Biologists turned to radar images, of all things, for part of their proof. Most of the eastern songbirds that migrate to the tropics take an overwater course across the Caribbean and the Gulf of Mexico, flying in great flocks high above the water. Weather radar can pick up the flocks, and radar images on file from the 1960s to the present show a clear diminishment of the autumn migration—a 50 to 75 percent reduction in that time.

In all, more than seventy-five species of Appalachian perching birds migrate to the New World, or neotropics, to winter. Some, like red-eyed vireos, travel all the way to the Amazon basin, but most concentrate in southern Mexico, Central America and the islands of the Caribbean—the region that has suffered the heaviest deforestation and habitat loss. Generally speaking, the migrants that have shown the worst reductions have been those like wood thrushes and olive-sided flycatchers that winter in mature tropical forest.

If songbird populations have declined as sharply as migration counts and diminished "waves" would suggest, then why haven't more people noticed equally sharp drops in the numbers of wood thrushes or orioles singing in their backyards? It may be because singing males make up only a small part of the overall songbird population; many more (perhaps even many times more) males cannot find a breeding territory, and so exist as a floating corps, waiting for an opportunity to occupy a territory. It appears that the population as a whole must drop significantly before there is a noticeable decrease in singing males. Poor observers that we are, we realize there is problem only when the normal safety margin of extra birds has been siphoned away to nothing. By then, the situation may be dire.

And, increasingly, people are seeing significant reductions in songbirds on the breeding grounds, not just in migration. The worst-hit areas have been in fragmented habitats such as urban and suburban parks, where forests stand isolated by cleared land. In the Washington, D.C., area some parks have lost several species entirely, and one recorded a 94 percent drop in the number of ovenbirds, once among its most common warblers.

The annual North American Breeding Bird Survey (BBS)—conducted at more than two thousand sites around the continent—has also shown alarming trends. Nearly three-quarters of the migrant, forest-dwelling songbirds in the northern Appalachians have experienced marked declines in the past several decades. Wood thrushes are down by 23 percent, eastern wood-pewees and northern orioles by a third, tanagers and various warblers by similarly drastic margins. During the decade ending in 1987, Tennessee warblers, which nest in New England and much of Canada, were dropping at a frightening rate of 11.6 percent per year.

This was especially surprising because the first fifteen years of the BBS showed most of these species slowly increasing—a result of abandoned farmland reverting to forest, it was thought. In the decade that followed those trends reversed, so that now many are falling at a much faster rate than they had been rising.

It is tempting to lay the blame for this problem at the feet of the tropical countries, which are cutting their forests with reckless abandon. But the riddle of why woodland songbirds are disappearing is far more complex than that. Even though the red-eyed vireo winters in western Amazonia, a region largely unaffected by habitat destruction until recently, it has decreased markedly in the East. Rose-breasted grosbeaks, which have dropped by roughly a third in the Northeast, can easily tolerate disturbed habitat in Central America.

No one is minimizing the impact of tropical deforestation; for birds that winter in mature forest, timbering and conversion to agriculture can be devastating. But more and more, experts are coming to see the vanishing songbirds as victims of multiple evils—a loss of wintering habitat combined with destruction of migratory corridors and rest stations, and a variety of troubles on the breeding grounds in North America.

It is not surprising that the worst-hit areas, like the urban parks mentioned earlier, are small patches of forest in a sea of nonwooded country. Generally speaking, the birds that have decreased or disappeared from these islands are those, such as black-and-white warblers and Acadian flycatchers, that prefer to nest in deep forests. Fragments offer less security, including the threat of heavy nest predation by raccoons, opossums, blue jays, grackles, crows, rat snakes and domestic cats—all animals found more commonly in disturbed habitats or the edge of woods than in large, intact forest.

Fragmented patches of woods also invite brown-headed cowbirds, North America's only true nest parasite. Once a plains species (and thus an anomaly in its otherwise tropical family), the cowbird female seeks out the nests of other birds, tosses the hosts' eggs and lays hers in their place. Many western grassland birds adapted to the cowbirds, recognizing and removing their eggs in turn.

Eastern songbirds had no instinctive defenses, however, and when the Appalachian forests fell and the cowbirds moved east, they found an abundance of naive hosts. In one study, fifteen wood thrush nests held just eleven thrush eggs—and nearly fifty laid by cowbirds. Other researchers have documented even higher parasitism rates, particularly in the Midwest.

Cowbirds are not a forest species, but they are quick to exploit any opening or avenue through an intact woodland. That's why fragmentation is such an important concept where songbirds are concerned. Pushing a road or a power line through a large area of forest does more than just cut it in two; because cowbirds and nest predators infiltrate along the openings, the "edge effect" (as it is known) from a single road can dramatically reduce or eliminate the area of safe, deep-woods habitat for warblers, thrushes and other migrant species. Take a five-hundred-acre tract of woods and scatter just a few houses through it—each with an access road, each with a bit of yard—and you may ruin it completely for tropical migrants.[3]

At a number of sites along the Appalachians, researchers have shown that the size of the forest plays a tremendous role in how successful the songbirds nesting in it will be. In New Hampshire's Hubbard Brook Experimental Forest—nearly eight thousand acres of intact northern hardwood forest—nine out of fourteen species were stable or increasing; surveys in large, relatively undisturbed areas in the Smokies and Connecticut also found stable or rising populations of many neotropical migrants. And in Pennsylvania, continuing research by Hawk Mountain Sanctuary has demonstrated the disadvantages of nesting in small fragments.

Shortly after dawn on a mid-June morning, I strode through the chestnut-oak forest with Jeff Hoover, a graduate student at Hawk Mountain. Hoover carried a battered old golf bag over his shoulder and was checking one of the study plots in a large, relatively intact area of woods along a ridge known as Owl's Head. "It looks pretty weird, but this is the easiest way to carry everything," he said, slinging the golf bag from his back and removing a series of long aluminum poles, stakes, a hammer and plastic bags containing mist nets.

I helped Hoover set up the long nets at right angles, forming an *L* thirty feet on each leg. At the inside corner he placed a tape recorder, and next to it a decoy—a carved and painted wood thrush perched stiffly on a twig. When he pressed the button on the tape player, the rolling, flutelike song of a thrush came booming out at several times normal volume. For a real male wood thrush defending this territory, such a challenge can't go unmet.

In less than a minute a male thrush appeared, hopping in obvious agitation in the trees over the decoy, but never coming low enough to become tangled in the fine meshed nets. "It's been like this for the last week," Hoover said quietly, watching the scene from a short distance away. "Ever since the females starting incubating, the level of aggression in the males has dropped way off. I haven't caught one in three days."

Since the wood thrushes arrived the previous month, Hoover had been trapping males, here and on other study sites, and marking them with colored leg bands so their movements could be traced; through binoculars, I could see that the bird displaying to the wooden decoy sported a series of red and silver bands. Hoover and other members of the research team were comparing deep-woods locations like Owl's Head with smaller fragments in the valleys to the east, to see if there were differences in nesting success, territory size and other factors.

As part of that project, Hoover was also looking for wood thrush nests and monitoring their progress. Following his pointing finger and careful directions, I eventually found the nest, a structure the size of a robin's built on a thin, horizontal branch about twelve feet off the ground. From its base dangled long streamers of grapevine bark, leaves and strands of grass, a trademark of wood thrush construction.

The female hunkered down so completely that I could see only the tip of her black beak protruding over the lip of the nest. Hoover quietly assembled another aluminium pole, this one with a round mirror mounted on an angled rod at the end. He eased the contraption up through the leaves until it was a foot above the nest; at this point the female finally bolted, scolding us from a branch nearby.

In the reflection we could clearly see a brownish, speckled cowbird egg nestled among three powder-blue thrush eggs. Hoover blew out a disgusted sigh and lowered the mirror. "Probably because of that," he said, gesturing to a tiny clearing in the woods about thirty yards away. "It doesn't take much with cowbirds."

Sure enough, when we walked past the small opening—barely fifty feet square—there was a male cowbird singing in one of the oaks along its edge.

Still, the larger and less fragmented the woods, the better a thrush's chances of bringing off a successful nest. Hawk Mountain's research has shown that in places like the Owl's Head plot, where forest covers several thousand acres, wood thrush nesting success was a comfortable 70 percent. That fell sharply as woodlots got smaller, however—a little over 50 percent of nests successfully raised young in two-hundred-acre fragments, and when the patches were fifty acres, the rate dropped to less than a third.

A related study focusing on ovenbirds, conducted on many of the same tracts as Hoover's wood thrush work, confirms the dangers posed by forest fragmentation. In unbroken woods of at least twenty-four thousand acres, nearly half of the ovenbird males holding territory successfully raised chicks. On fragments—including surprisingly large, five-hundred-acre blocks of forest—the success rate dropped to almost nothing, even though all but the tiniest fragments had a high density of ovenbirds holding territories. Overall, among both wood thrushes and ovenbirds, cowbird parasitism did not seem to be much of a factor, leading the research team to conclude that "edge" predators may be the biggest threat.

Other biologists have made similar findings. A study comparing nest predation in the Great Smokies with small fragments in Maryland showed that only 2 percent of the nests were destroyed in unbroken forest, versus nearly 70 percent in suburban woodlots.

To a casual observer, these small fragments seem to have healthy wood thrush or ovenbird populations; when you walk through the woods on a May morning, you hear their calls ringing from every direction. But they may in fact be acting as biological black holes, attracting birds but allowing them little chance to breed successfully and so weakening the species' chances overall.

Neotropical migrants may be especially vulnerable to predators and cowbirds, which could explain why they are declining while most resident species, such as chickadees and cardinals, are stable or increasing. Tropical migrants tend to build open cup nests that are more exposed to danger, and because they travel such a long distance to get here, they usually have time to raise only one brood. Resident birds, on the other hand, needn't expend so much energy in migration and can get a head start on the nesting season, routinely raising two or three batches of young. Many also choose more protected nest sites, including tree cavities.

Ornithologists agree that saving the neotropical migrants may be the toughest conservation challenge we've ever faced. Instead of one easily identified and easily targeted cause—DDT use or the burning of high-sulfur coal—there are dozens, from timbering practices in the northern Appalachians to housing construction in the Piedmont and the loss of forest to banana plantations in Central America, problems that stretch across thousands of miles and dozens of countries. Each species of songbird probably has its own cadre of troubles (the Tennessee warbler, for instance, may be in such a steep decline because of a drop in spruce budworm, a favorite food), making a search for panaceas useless.

Worrisome as the situation is, there is still time to turn things around. Songbirds are declining, but even those like the wood thrush that are showing the largest decreases still number in the millions, and only a few are in imminent danger. Such a broad-based problem calls for broad-based solutions, and over the past two decades, a number of comprehensive initiatives have been launched, including Partners in Flight—Aves de las Americas, a hemispheric effort involving governments, private conservation organizations and individuals.

Several days after my visit with Jeff Hoover, I tagged along with a friend who was leading an early morning bird walk for grade school–age kids. There was plenty of giggling and a little roughhousing—this was summer vacation, after all—but when someone spotted a bird, the group fell quiet and looked up. Binoculars wavered around inexpertly, but soon most of them found the spot.

High in the branches a cluster of leaves jostled, and the most brilliant scarlet tanager imaginable hopped out into the open. Almost in unison, the children gasped—the tanager was fantastically red, its black wings and tail the perfect counterpoint. There would be time later to explain to the kids about the hazards this tanager faces and its uncertain future. For now we simply let them drink in the sight, knowing that another generation was learning that there are some things in the world worth fighting for.

American Golden-plover

Chapter 11

ULTIMA THULE

To the Greeks and the Romans, the land at the highest latitudes was *ultima Thule*, the uttermost north, a place of cold and ice and legend.

The Appalachians have an ultima Thule of their own, lost in mist and uncertainty, in the ice-choked waters between Newfoundland and Labrador. Here, fifteen miles from land, the last of the ancient hills rises above the sea like a breaching whale, glacier sculpted and barren, the home of Arctic foxes and seabirds and little else: Belle Isle.

In a straight line, it is more than twenty-one hundred miles from the fertile slopes of Cheaha Mountain in Alabama to the cliffs of Belle Isle, but it might as well be twenty thousand miles, or two hundred thousand, for this is a world apart. None of the conventional notions of "Appalachian" have any meaning here, for Belle Isle is Arctic to the core.

The boat pitched hard as Bill Carpenter peered through the fogged window of the forty-eight-foot *Trudine Norman*, trying to see through the murk outside. The deck was pitching in a heavy swell, and the whitecaps looked uncomfortably like ice "growlers," the dangerous chunks from an iceberg's final decay that can punch a hole in a boat as surely as striking a rock. Unlike their parent bergs, growlers sit too low among the waves to show up on the green radarscope at Carpenter's elbow, and we were all a little on edge.

The weather forecast, so far as it went, had been right; the winds were light, and Carpenter thought he'd be able to make the tricky run to Belle Isle and the even trickier task of getting me ashore. There is no harbor anywhere along the nine-by-five-mile island, only a small cove near one of the two lighthouse stations, and I would have to be shuttled in by rowboat and then scramble up the concrete landing—if the waves weren't too rough, that is.

But the clear skies had given way to dense fog and, although the surface winds were calm, a hard blow the day before had left a legacy of heavy swells rolling in from the southwest. We were only two miles out of harbor, and the seas were getting worse by the minute.

"If it's like this out at Belle Isle, there's no way we're going to get you ashore," Carpenter said around his cigarette, not taking his eyes off the water ahead. A vicious one-two punch from the waves sent much of my gear

Dropped by the glaciers that scraped the island bare, a huge boulder sits amid the treeless tundra of Belle Isle, the northernmost point on the long Appalachian chain.

tumbling across the cabin, and I saved myself from falling only by grabbing the chart table. "That's it," he said flatly, spinning the wheel. "There's no sense in going out there. We'll try again later if it calms down."

The Strait of Belle Isle, the narrow strip of ocean separating western Newfoundland from Labrador, has some of the worst—and least predictable—weather in the world. A clockwise ocean current circling Newfoundland pushes north through the straits, while a frigid tongue of water dropping down along the Labrador coast passes it going the other way. Warm water and cold roughhouse through the passage, with Belle Isle sitting square in the middle of the melee. It was now August, and the weather would only deteriorate as fall approached.

We passed the rest of the day in deceptive sunshine, sitting on the wharf in the tiny village of Quirpon shucking freshly caught snow crabs, sucking down the cool, white meat and tossing the scraps to the fish. The fog bank stayed just beyond the harbor, and the lightkeeper on Belle Isle radioed us periodically with the dreary news that conditions were actually worsening. On a clear day, you'd be able to see the island from Quirpon (the name, strangely enough, rhymes with "harpoon"). But the fog kept it sealed and secret, and I began to wonder if I'd ever see the Appalachians' last hurrah, much less walk its hills.

I'd wanted to visit Belle Isle for years, ever since I came across mention of it in a geology book, the northernmost point on the Appalachians. But aside from locating it on maps, a lance-head sliver off the Northern Peninsula, I had been unable to discover anything about it at all—its topography and ecology, whether it was inhabited, if it would even be possible to go there at all. Even the Newfoundland tourism department was little help initially—the first person to whom I spoke confused Belle Isle with Bell Island, a short distance from the capital of St. John's.

"Oh, yes," she chirped, "there's a ferry service that runs twice a day and only costs two dollars." Looking at the map, at the speck of land so far out in the ocean, I knew she couldn't possibly be right.

The reaction near Quirpon was different, but no less enlightening. I'd tell people I was going out to Belle Isle, and they would stare for a moment, then blurt out, "Why?" A young woman in a grocery store flashed me a grin and nudged my side when I told her, an "Oh, go on" sort of gesture, as though I were making a joke at her expense. No, really, I told her. Her smile faded, replaced by a look of worried bewilderment. Even in normal times only fishermen went to Belle Isle, and with the collapse of the cod fishery, no one, it seemed, went there at all anymore. Maybe she thought I was a little bit crazy.

Maybe I was. The night before we set out, I'd sat in Bill Carpenter's living room in St. Anthony, just south of Quirpon, looking at snapshots of Belle Isle. Carpenter, a math teacher in town, passed me another photo album and flipped to the pages of the black cliffs with white breakers at their bases. "It's an unfriendly place," he said. "There's no secure harbor—it looks like somebody just pushed it up from the sea, right out of deep water. It's windy, treeless, just covered with what we call blackberry barrens. Not a place you want to be caught in bad weather." He flipped another page, to photos he'd taken from a friend's airplane the previous August. The hills still held pockets of snow, their runoff feeding the more than three hundred ponds that dot the island. The pictures showed a rugged, brown landscape that hardly looked inviting.

All that next day, after our aborted passage, we sat in the Quirpon harbor waiting for the weather to clear, and the talk kept returning to Belle Isle. Locals who wandered down for a chat (and perhaps to find out why anyone would come so far to visit so desolate a place) added their two cents' worth. One fellow mentioned a cave in the sea cliffs that the French fishermen, who had exclusive rights to the west coast of Newfoundland until the late 1800s, used as a morgue. When one of their own died, he said, they'd salt him down, stick him in the cave and load him up with the cargo when it was time to sail home.

Another mentioned the cache of aluminum poles found on the island a few years back, perhaps part of a German communications system in World War II, when the Strait of Belle Isle was a hunting ground for U-boats, and when German submariners would occasionally land to commandeer diesel fuel. As a consequence, the inhabitants were ordered to bury their drums of oil back in the White Hills beyond town.

Several people mentioned the madwoman of Belle Isle, although no one could agree on the details. One version styled her a young woman of some note who became pregnant by a cabin boy. Her enraged father had her marooned—with the lad and her maid—on Belle Isle. By next spring the cabin boy and the maid were dead, and the woman was mad. Little wonder Belle Isle was also known as the "Isle of Demons." The voices flowed with the accent peculiar to Newfoundland, a blend of archaic British and Irish with aspects of Scots and French pronunciation for seasoning. It is a rapid-fire, clipped speech, in which the bottoms drop out of some words and hit an inexplicable bump in others; father becomes "fadder," ice becomes "oyice" and Belle Isle is "B'loil," a single word.

The weather outwaited us, and by early evening we called it quits for the day, since there was no longer time for the boat to make it back to shore in daylight, when the ice is easier to avoid. When, the following day, the seas at last subsided and the wind cooperated, we sprinted across the strait on oil-flat water, fulmars and shearwaters riding the breeze before us, puffins and murres and guillemots buzzing across the bow on frantic wings. The fog hung just a few yards above the surface, but it lifted as we cleared land, and two hours later, through gaps in the mist, I could finally see Belle Isle.

With no trees to soften its contours, the island makes a foreboding first impression. Cliffs shoot up from thirty or forty fathoms of water and end in rugged, jumbled peaks that crowd together. The cove where I was to land scarcely merited the name, a mere indentation in an otherwise sheer drop, with a set of old concrete steps ending in a landing about six feet above the waterline.

I could see figures standing at the top of the steps, waving: Randy Campbell, the lighthouse keeper, and his wife Emily, who was taking advantage of the rare boat trip to make a short visit to her family on shore. Randy turned out to be an affable man of thirty-five, blond hair whipping in the freshening wind, a clipped mustache, weathered hands. He waved good-bye to his wife, who sat in the rowboat surrounded by boxes filled with crochet work, destined for a craft store in St. Anthony.

We loaded my gear onto a tractor for the long, steep ascent to the lighthouse, and a short time later I set out over the last hills of the Appalachians.

When the lark stopped moving, it blended so perfectly with the rocks around it that, even with ten-power binoculars, I had a difficult time distinguishing it. Then the wind tickled one of the tiny feather "horns," and the whole bird snapped into focus.

I'd been walking the highlands of Belle Isle for hours, and the pair of horned larks were the first land animals I'd seen, not counting the Mongol horde of black flies that even the cold wind could not abate. The birds were perched near the top of a low crag, one carrying a beakful of caterpillars, so I knew they had a late nest somewhere nearby.

The larks and I sat along the edge of a great bowl in the hills, rimmed with bare knuckles of rock too low to be mountains but too imposing to be called anything else. Below were several ponds, ranging in size from tiny mirrors I could jump across to one that may have covered thirty acres. The flies aside, I was exquisitely comfortable, cushioned by the thick mat of tundra vegetation on which I sat, a natural mattress of surpassing softness.

It was also a mattress of surpassing beauty, for the tundra was in a full bloom; the tapestry of wildflowers extended on every hand with an amazing abundance and variety. Labrador tea, dwarfed to just a couple of inches, raised fluffy balls of white, mingling with the pink of pale laurel and twin-flower and the blue trumpets of butterwort, whose sticky leaves serve to trap and digest tiny insects. Swedish bunchberry, a small, dark-centered European species also found around the Gulf of Lawrence, was everywhere, a reminder of the time when the northern hemisphere was a seamless whole.

Many plants, like three-toothed cinquefoil, mountain sandwort and *Diapensia*, formed dense mats, while along the boggy edges of the streams and ponds, lush stands of blue flag iris, leafy white orchis, cotton grass, cow parsnip and fireweed grew in profusion. On the hillsides, a few spruce managed to survive by spreading out across the ground: two-dimensional trees. Many of the species were the same as I'd found on top of Camel's Hump and other New England peaks—only here they were growing within a few hundred feet of sea level.

The growing season this far north is excruciatingly short—the last killing frost is usually in late June and the first of autumn is sometime in mid-September, although frost may occur in any month of the year. Consequently, spring, summer and fall crunch together like cars in a freeway pile-up, and the resulting bloom is explosive and spectacular. It is also unsettling to one from more southerly latitudes; in St. Anthony, I was taken aback to see fireweed and goldenrod, both autumn flowers, blooming side-by-side with columbine and lilacs, early spring blossoms where I live. Nor is the soil thick or fertile, compounding the difficulties for plants. On the main island of Newfoundland, you'll often see vegetable patches growing—miles from any visible residence—right along the berm of the road. The placement isn't for the sake of convenience, but because the road construction scraped up what little dirt there was and piled it along the edge. The berm is the only place with enough soil to support a potato.

As with alpine tundra on more southerly mountains, true Arctic (and subarctic) tundra traffics heavily in berries, and Belle Isle is no exception. The most revered hereabouts is cloudberry, or "bakeapple," a nickname easily explained by one taste of the large, orangish berry, which is borne singly and upright above a cluster of leaves.[1] Randy and his wife, like many Newfoundlanders, work hard during the bakeapple season, laying up great quantities of frozen or canned berries. This season's crop was poor and was anyway not quite ripe in this, the second week of August, so he treated me to some of last year's—a bowlful of delicious, yellow-orange berries swimming in their juices, which looked and tasted a great deal like spiced, baked apples.

Cloudberries get the glory, but the most common fruiting plant on the tundra is crowberry, whose mats of glossy, evergreen leaves cover much of Arctic and subarctic Canada, and which bear a wealth of small, purple-black berries. Newfoundlanders know them as "blackberries," and while they pick them, it is not with the same fervor as for bakeapples. But if crowberries are slighted by humans, wild animals have long appreciated them. In fact, crow-

The white flowers of cloudberry—known in Newfoundland as "bakeapple," and beloved for its orange, raspberry-like fruit—bob in the breeze on Belle Isle.

berries once played an absolutely vital role in the hemispheric migrations of the Eskimo curlew, one of the most abundant birds on the planet.

Except for birders, few people have heard of an Eskimo curlew, even though its numbers may once have rivaled those of the passenger pigeon or the bison. At fourteen inches the smallest of the North American sandpipers known as curlews, it had the group's characteristic long, down-curved bill, brownish plumage with a dark cap and light eye-stripes and a rich wash of cinnamon beneath each wing.

Eskimo curlews nested in the Arctic and wintered on the South American pampas, making a circular migration across the Andes, up through Central America and the Great Plains to the Northwest Territories; in fall they flew east to the coast of Labrador and Newfoundland and over the open Atlantic to southern Brazil and Argentina. Just before making the final, twenty-five-hundred-mile vault over the sea, the curlews would feed voraciously on crowberries, filling their crops again and again, staining their breasts with purple juice. They laid on fat lavishly, so that a single bird might weigh as much as a pound.

Their numbers beggared belief. I can stretch my imagination and accept the vast flocks of passenger pigeons, for the hardwood forests of the central Appalachians are so obviously fertile, but the thought of tens of millions of curlews descending each August from the barren Arctic is much harder to believe—and yet it happened. "The accounts given of these Curlews border on miraculous," John James Audubon wrote in 1833, when he was exploring the nearby coast of Labrador. A few weeks later, when the migration had started, he saw for himself.

"They fly in compact bodies, with beautiful evolutions, overlooking a great extent of country ere they make a choice of a spot on which to alight; this is done wherever a certain berry, called here 'Curlew berry,' proves to be abundant ... in an instant all the ripe berries on the plant are plucked and swallowed, and the whole country is cleared of these berries as our Western woods are of the mast."

Such numbers, such numbers; one report from the Labrador coast in the 1860s tells of a curlew flock a mile long and a mile wide, while in Newfoundland the flocks blotted out the sun and made a noise—the high, whispering flight notes of millions of curlews—that sounded from a distance like wind blowing in the rigging of a great ship or the far-off jingling of sleigh bells. A schooner captain in 1851 near Bermuda reported a curlew flock that passed him for two days and a night, calling all the time. (Interestingly, some experts believe the "river of birds" that Columbus' flotilla intercepted on October 7, 1492, may have been largely curlews. The sighting convinced his mutinous crew that land was near, and they followed the birds, hitting a Caribbean island four days later.)

The curlews could not last, not with the immense pressure that humans applied to them, the same sort of assault brought to bear on the other super-abundant species of North America. Curlews were shot at every opportunity, especially in Atlantic Canada and New England on their southern migration, after they'd fattened nicely; in one day on Nantucket Island in Massachusetts, seven thousand were killed. They were known as "doughbirds," so laden with fat that their tightly stretched skin was said to tear open on impact with the ground. People ate many fresh and canned the surplus. The batteries of guns awaited them at every turn, here and on their South American wintering grounds, and particularly on the American prairies each spring, where they provided easy meat after a long winter. Curlews fell by the wagonful.

There may have been other evils at work, some caused by humans, some not. In spring, while passing through the prairies, curlews fed heavily on grasshopper egg cases, a food supply that probably decreased as the native grasslands were converted to crops. Other authorities implicate low-er-than-average summer temperatures in the 1880s, a result of global vol-canic explosions, and a series of severe autumn storms that may have dis-rupted the migration. (Others, however, studying tree rings in the Yukon, have pointed out that the 1880s were actually warmer than usual in parts of the Arctic.) But even if these factors had an impact on the curlews, simple slaughter had the greatest effect.

The curlew population began a noticeable downturn in the 1870s and collapsed before the turn of the twentieth century; observers in Labrador reported them missing entirely by about 1890. The last one shot in New England was killed in Massachusetts in 1913, and in 1929 the Eskimo curlew was declared extinct.

Reports of its demise were premature, however. A curlew was shot in Labrador in 1932, and sightings have dribbled in ever since—a bird here and there, mostly in spring along the Texas coast, almost always alone. There hasn't been a confirmed report—that is, one backed up by a specimen or clear, unequivocal photographs—since a curlew was shot in Barbados in 1963. The last sighting that the U.S. Fish and Wildlife Service considers reli-able was in Nebraska in 1987, and birders acknowledge that Eskimo curlews are easy to confuse with other species, including the very similar whimbrel.

Because shorebirds are among the most long-lived of birds, this pro-longed series of sightings is not as hopeful as it might seem; some experts have suggested it was mostly the same, increasingly geriatric birds year after year, with nothing to prove they were even breeding. The Arctic is, after all, a vast region, and the handful of surviving curlews might be so widely scattered across it that they wouldn't be able to find each other to mate.

A four-year search of the breeding grounds in the northwest Canadian Arctic failed to find any curlews in the early 1980s, although two were reported in the Northwest Territories in 1987, in a region where they once nested abundantly. As sightings continue to dribble in every few years, it seems possible (if increasingly unlikely) that at least a few curlews are finding each other out there, somewhere. While that is something very much to be hoped for, it is nevertheless a pathetic comedown for a bird that once could make the Newfoundland sky sing with the sound of celestial sleigh bells.

The curlews were never far from my thoughts on Belle Isle, and more than once I daydreamed about finding one on a lonely hilltop fen. I thought of them when, coming over a rise to a small pond, I scared up a half-dozen fast-flying birds. They were obviously not curlews, with the short necks and beaks of plovers and the breathtaking speed of these tireless migrants. But

The looming cliffs of the Long Range Mountains form a grim wall near Western Brook Pond in Gros Morne National Park.

when they wheeled around in the sun, I caught a flash of brass on their backs and felt a surge of excitement. They were American golden-plovers, a species of shorebird that nests in the High Arctic and migrates south over the open ocean, only rarely showing up on land; its migration route parallels that of the vanished curlews. The flock spun by again, shining metallically in the high sun, then sank below the lip of the hill. Like the curlews, the plovers feed heavily on crowberries, and like the curlews, these once-abundant birds were devastated by the market hunting of the nineteenth century. They fared somewhat better, although now, more than a century later, their numbers remain just a shadow of what they once had been.

At times, I felt the empty hills of Belle Isle were crowded with ghosts—not the demons of the old stories, but those of vanished animals. The curlews haunted me most, for this would have been the peak of their migration season, but there were other spirits. In pre-European days, the flightless great auk—a penguin-sized bird that occurred only in the north Atlantic—migrated through the strait and nested on islands north and east of Newfoundland. The big, clumsy birds were hunted by Natives (a cache of two hundred beaks was found in an Archaic Maritime Indian grave in Port au Choix, a hundred miles to the south) and even more ruthlessly by the Europeans, who clubbed them for meat and sacked their nests for eggs. At times they could be herded right up gangplanks and into the ship's hold, fresh meat on the hoof, as it were. Great auks were last seen in Funk Island off the northeast Newfoundland coast in 1852, another loss.

Nor were the ghosts all birds. All across the island, for instance, are the faint but discernible trails of caribou, thin pathways through the turf. They look recently used, but this is a result of the extraordinarily slow growth rate of tundra vegetation. In truth, the caribou have been gone for more than a century, shot out by fishermen.

Today, the largest mammals to reside permanently on Belle Isle are red foxes, one of which comes in nightly to the light station for the scraps placed out for it by Maude Elliott, wife of Carl, the assistant lightkeeper. Some years there are Arctic foxes on the island as well, but they come and go over the winter pack ice, and this year there were none to be seen.

I came across scat from the red foxes often in my wanderings; it was invariably made up of feathers and bird bones (the result of scavenging at the seabird colonies on the cliffs), the fragments of crab shells and silver flecks of fish scales. It is difficult to imagine a harder, less forgiving place for a predator to make a living.

There was also evidence of far more formidable hunters. Scrambling up the side of a high hill, more than six hundred feet above the ocean,

I spotted something white in a crevice beneath a lichen-crusted boulder. Reaching down, I pulled out a skull—blunt, big-eyed, with wide cheekbones like a cat's. In another, forested place I would have guessed a lynx, but there are no cats on Belle Isle, and it only slowly dawned on me that it was the skull of a young seal.

The waters of the Strait of Belle Isle—and the whole of Newfoundland and the Labrador coast, for that matter—are rich with seals. On the way out to the island, we saw them poking their doglike heads above the water, staring like startled pedestrians, before dropping under with an explosion of bubbles. There are five species found here, including the enormous gray seals and hooded seals, but the most common are harp seals, which each February and March haul out onto the pack ice by the hundreds of thousands to bear their young. This was the skull of such a pup, and while it was possible a fox had tugged parts of the thirty-pound carcass up this high, I think it much more likely a polar bear had carried the dead seal up here to eat.

Polar bears are the supreme hunters of the Arctic ice pack, and each winter a few follow the freezing ocean south to Belle Isle and Newfoundland's Northern Peninsula, hunting seals. These are the biggest bears in the world, with one on record that weighed 2,210 pounds and measured more than 11 feet long; most are smaller, with about a thousand pounds being the average maximum for a male. Nevertheless, that is awesomely big for any land mammal, to say nothing of a predator. There is something about polar bears that I find at once compelling and terrifying—a visceral, understandable reaction to a carnivore so big that it has been known to kill small whales.

The sea ice starts to form off the northern coast of Newfoundland in December and seals off most of the Northern Peninsula for winter. With it come the bears, moving south from the Arctic on migrations that can be surprisingly lengthy; some bears are known to have traveled from the islands of northern Hudson Bay to southern Labrador, a distance of more than six hundred miles. The pack ice, bullied by wind and water currents, fractures and shatters, creating open-water passages and a realm of shifting, crumpling ice floes. The great bears are completely at home in this transient world that is neither land nor ocean. They swim superbly (polar bears have been seen paddling contentedly nearly forty miles from the nearest land), and will often plunge beneath the ice for long distances, using the breathing holes of seals when they need to surface.

One could not design a hunter better adapted or equipped for the subarctic winter. Polar bears have large, flat forefeet, which function equally well as swimming paddles and snowshoes; all four feet are covered with thick layers of insulating hair. The body fur, although not the dense insulator of some mammalian coats (a heavy blubber layer takes that role), plays a unique role in keeping the bear warm. The hairs are hollow and translucent, channeling solar radiation directly to the skin—which is in turn black, allowing for maximum absorption of the feeble northern sun. Polar bears even have a specialized nictitating membrane in each eye that prevents snow blindness. Inuit hunters traditionally achieved the same effect by carving slits in a pair of bone goggles.

Polar bears probably have the least contact with humans of any species of bear in the world, and this may be why they have always been considered so dangerous. Without any ingrained fear of mankind, they are perhaps more apt than other bears to view humans as just another meal. This is far from universal behavior, however, and most that do rub shoulders with people, like the polar bears that gather each autumn in Churchill, Manitoba, generally leave the humans alone.

The same seems to be true along the Strait of Belle Isle. The subject of polar bears came up one day in Quirpon. "Oh yes, we see them here in the winter," one fellow said, "but they usually aren't a problem. We leave them alone, and they leave us alone."

Sometimes circumstances force a different outcome. A few years back, a polar bear stayed on Belle Isle long after the pack ice broke up and made itself a disturbingly regular presence near one of the two light stations. Such a situation was simply too dangerous to tolerate, and the bear had to be shot by the authorities.

Polar bears are only seasonal visitors to Newfoundland, and they are no doubt less common there than in times past; when John Cabot first saw the island in 1497, he wrote that it was "crowded with white bears." If the great hunters ever lived year-round on Newfoundland, they have not done so since Europeans came to stay.

My first afternoon on Belle Isle, Randy took me far up to the center of the island, where a military plane crash-landed in the 1940s. As we were picking through the widely scattered wreckage, I came across several white bones poking up from the vegetation. They were long and heavy, one almost as thick as my forearm.

"Polar bear bones," Randy said matter-of-factly. "We find them, sometimes."

In spite of myself I looked over my shoulder, but we were alone. Yet it took a few moments more for the hairs on the back of my neck to settle down again, and that night, as I lay on the dreaming fringe of sleep, I could see a white bear stalking the wind-raked hills of a land at once foreign and familiar.

Wandering north through the Appalachians is an exercise in reductivist ecology, with the southern cove forests at the fat end of the diversity curve and northern Newfoundland the sparse minimum.

The Great Smokies, for example, have more than two hundred species of breeding birds, sixty species of mammals, twelve kinds of frogs and toads, twenty-five species of salamanders and twenty-one species of snakes. The Long Range Mountains in Newfoundland, on the other hand, have about a hundred bird species, just thirteen native land mammals including two bats (a fourteenth, the gray wolf, is extinct),[2] three frogs and a toad, and no snakes or salamanders at all.

The extreme paucity is due partly to the predictable falloff of species the farther from the equator one travels (a function of climate, for the most part), and partly to Newfoundland's being an island. Insular ecosystems are

Mountains meet the sea near Lobster Head Cove in Gros Morne, where a lighthouse guards the point.

almost always less diverse than neighboring mainland ecosystems, and for obvious reasons—the animals and plants must cross a barrier to reach the island, and many cannot do so.

Biologists studying Newfoundland's mammals have tried to figure out where they came from and when they arrived. Among the most significant clues are the similarities between the forms living on the island and populations found on the mainland. The otters and meadow voles found on Newfoundland, for example, seem to be most closely allied with those in the Maritimes, suggesting that they swam (or, in the case of the voles, rafted on vegetation) across the Cabot Strait.

Newfoundland black bears, on the other hand, are thought to have colonized the island from Labrador across the much narrower Strait of Belle Isle, perhaps prior to the last glacial advance; if so, they must have survived in the scattered, ice-free refugia along the coast known as nunataks, much as black bears do today in glacial areas of Alaskan coast. The length of isolation brought on changes, however, as isolation usually does; the Newfoundland bear has a distinctly higher skull and shorter snout than its mainland relatives.

This raises another important point about island species—they tend to be different from those on the mainland. Sometimes the differences are slight, as with Newfoundland's fauna. While many of the mammals there are noticeably different from their mainland cousins—the meadow vole is quite pale, the marten larger and darker than normal—none warrant status as full, separate species. Given enough time by itself, an island usually evolves a unique mix of animals and plants found nowhere else. The process has only just started in Newfoundland.

Other mammals, among them Arctic hares and caribou, are essentially identical to mainland forms (so too, apparently, was the now-extinct wolf). Since hares, wolves and caribou can and do cross large areas of ice, especially when populations are unusually high, enough may have filtered back and forth from the mainland during the glacial maximum, and off and on since then on winter pack ice, to negate the evolutionary effects of isolation.[3]

To the original tally of fourteen native mammals, humans have added another thirteen to the island in historic times. The reasons for making the introductions may seem ludicrous (chipmunks were stocked in one provincial park in the 1960s for "aesthetic reasons"), while others show at least a tinge of forethought, like the release of masked shrews in the hope of countering a larch sawfly outbreak.

Two of the most successful introductions were those of the snowshoe hare in the 1860s and moose around the turn of the last century. Moose are now so common in the forested regions of the province that they are a significant traffic hazard (one popular poster shows a picture of a moose with the words "Newfoundland Speed Bump"). Driving to St. Anthony late one day, after eight tiring hours on the road, I spotted a young bull standing by the edge of the trees. Normally in moose country I drive slowly and with an alert eye, but my concentration had lapsed, and I was traveling rather faster than I should have been. Even as I stomped for the brake, the moose bolted, charging up the steep embankment toward the road. A collision seemed inevitable, but at the last moment the moose veered away from me, running along the edge of the road while my car squealed to a belated stop. For long moments I could do nothing but draw shaky breaths, mesmerized by the carnage we narrowly avoided. When a moose and a car collide, the moose has the better chance of walking away from it.

Snowshoe hares have also adapted well to their new home, but—like any introduction, however seemingly innocuous—their presence has had far-reaching effects on Newfoundland's ecosystem.

The sudden abundance of hares provided a new prey species for the lynx, which was originally rather rare on Newfoundland; the number of lynx blossomed as a result. Unfortunately for the lynx, snowshoe hares undergo huge boom-and-bust population cycles, normally on an eight- or nine-year rotation. At the cycle's peak, the hares are everywhere; at its nadir, they all but vanish. When the numbers of hares collapsed, the abnormally high lynx population has to turn to other food sources, including caribou calves and ptarmigan.

The lynx's greatest impact, though, was on another native mammal, the Arctic hare. Weighing up to twelve pounds, it is North America's largest rabbit and inhabits the Arctic coast and islands from the Northwest Territories to Labrador; Newfoundland is the most southerly extension of its range. The Arctic hare is well suited to the cold—the extra body mass provides insulation, and the extremities, including the ears, are short, keeping heat loss to a minimum. Like the snowshoe it turns white in winter, and at higher latitudes remains white all year. More social than most rabbits, the Arctic gathers in large groups in spring and summer, and males may box over females. They often stand upright to look for danger and may run that way for short distances. They've even been known to scavenge the carcasses of other animals for meat. They are, in all, most unharelike hares.

At the time of European colonization, Arctic hares were common over much of Newfoundland, especially at higher elevations. The newly introduced snowshoe hares competed with the Arctics for food, however, and even worse, the inflated lynx population that fed on them—suddenly deprived of food by the periodic collapse of the snowshoes—turned to Arctic hares in the lean years. During the first three major snowshoe hare cycles, from the 1890s through 1920, Arctic hare numbers dropped drastically. Today they are restricted to only a few pockets in the Long Range Mountains, including Gros Morne National Park, where the treeless, windswept mountaintops provide some refuge.

Hiking in the spruce forests of the Long Range, I often came across heaps of cone scales, the hard, overlapping pieces that shield the spruce's seeds. Some of the piles were small, but many were quite large, middens that rose a foot above the ground and spanned three or four feet, the detritus of hundreds of shucked cones.

They were the work of red squirrels, a common species across the boreal forest but another mammal not native to Newfoundland. They were brought to the island in the 1960s to provide food for pine martens, a valuable furbearer that had been overtrapped (martens, which elsewhere feed largely on squirrels, originally preyed on voles on Newfoundland). Freed in a land full of spruces, the red squirrels rapidly spread, although the hoped-for increase in martens did not follow. More importantly, the red squirrels have come to so dominate the annual spruce cone crop that a native bird, the Newfoundland race of the red crossbill, is vanishing. The story is a fascinating and instructive one, mingling evolution, ecology and the law of unintended consequences.

Until about nine thousand years ago, Newfoundland was almost completely sealed in ice. When the glaciers retreated, the island was recolonized by spruces (which may have survived in the few ice-free pockets along the west coast) but not by red squirrels, which presumably could find no way across the Cabot or Belle Isle straits. Birds, however, could cross the distance easily, and so the island became home to two species of crossbills, the white-winged and the red.

Crossbills have one of the most peculiar bill designs in the avian world. The mandibles are massive at the base but taper to long, thin tips, which, instead of meeting neatly like almost every other bird's, twist around each other. Odd as it may appear, the crossbill's beak serves a very particular purpose—the opening of conifer cones. By sliding the tips of the bill between the scales of a cone and opening the beak, the crossbill levers the scales apart, allowing its tongue to reach in and remove the kernel. It is a neat and efficient method of extracting the nutritious seeds. Smaller cones are often removed from the tree and held in the crossbill's strong feet—left foot or right, depending on which direction the crossbill's lower mandible points. In other words, crossbills may be "left-handed" or "right-handed."

Ornithologists have long recognized that bill size in crossbills varies from region to region, even within the same species. Red crossbills, *Loxia curvirostra*, which are found in conifer woods across the northern hemisphere, range from relatively tiny-billed races on mainland North America to large-billed European birds. Scottish crossbills, until recently considered a red crossbill subspecies, have even heavier bills, and the parrot crossbill of Scandinavia has an almost grotesquely big beak.

The Newfoundland crossbill, *Loxia curvirostra percna*, is endemic to the island—that is, it is found nowhere else. It has a larger bill than its mainland cousins, which is puzzling, since *percna* feeds mostly on the seeds of black spruce, a thinner-scaled species than the white spruce. Biologist Craig Benkman, who has made an intensive study of North America's crossbills, figured out the riddle—and it comes down to squirrels. In the nine thousand years since the glacial retreat, and in the absence of red squirrels, the scales of Newfoundland's black spruces thinned considerably, so that today they average about 15 percent thinner than scales on mainland black spruces. (Benkman theorizes that the spruces reduced scale thickness as a way of channeling more energy into seed production.) Where squirrels occur this strategy would be disastrous, for the thin cones would be even easier prey, but on Newfoundland it worked well.

But if the scales are thinner, why would Newfoundland crossbills need heavier bills than mainland birds? Unlike red crossbills on the mainland and the white-winged crossbills that also inhabit Newfoundland, *percna* crossbills are not nomadic. They do not wander widely, following the heaviest cone crop, but stay on the island year-round. That, Benkman says, means that they must be able to feed on the closed cones of black spruce—unlike the thinner-billed white-winged crossbill, which prefers slightly opened cones. When the local supply is depleted, the whitewings move on, an option not available to the sedentary red crossbills, which must turn to the large percentage of black spruce cones that remain closed all winter. Hence the bigger, more powerful bill.

Newfoundland crossbills tread a very fine line. If the black spruce scales were any thicker, the birds would not be able to open enough, quickly enough, to avoid starvation. As it is, Benkman's research indicates that a crossbill needs to eat a spruce kernel about every seven seconds, all day, to stay alive through winter.

Enter the red squirrels, which were stocked in western Newfoundland starting in 1963. The island must have been a paradise, what with thousands of square miles of spruces, all bearing cones with exceptionally weak protection. The squirrels are now common on the main island of Newfoundland, and at densities markedly higher than on mainland.

The effects on red crossbills were immediate and devastating. With the squirrels monopolizing the spruce crop—they remove up to 90 percent, according to one study—the crossbills were robbed of their only steady winter food supply. The number of red crossbills plummeted in the early 1970s, and the species has all but vanished from most areas where it had formerly been rather common. The occurrence of nomadic white-winged crossbills on Newfoundland has also dropped, doubtless for the same reason. The only hope for Newfoundland crossbills may be the even smaller islands that ring the main island, some of which have not yet been colonized by squirrels; Anticosti Island in the Gulf of St. Lawrence also provides a squirrel-less refuge.

Benkman theorizes that, even if the crossbill can withstand the competition of red squirrels, the black spruce will be forced to evolve thicker scales, as spruces on the mainland already bear. That would also drive the Newfoundland crossbill to extinction unless it evolves a still heavier bill to cope with the tree's heightened defenses. But considering how swiftly and completely crossbill populations have crashed in the presence of squirrels, the point may be strictly academic.

Ironically, even mainland red crossbills haven't escaped the consequences of human actions. The subspecies of red crossbill—*L.c. neogaea*—once found in New England, the Adirondacks and the Great Lakes region, which was adapted for feeding on the cones of the virgin white pine stands, was virtually eliminated when the pineries fell for timber. Its disappearance was masked by the irruptive nature of other crossbill populations, which still "invade" the region every few winters, and by the expansion eastward of yet another subspecies, *L.c. bendirei* of the Rockies. It wasn't until old specimens from collections dating to the turn of the century were reexamined in the late 1980s that the loss became clear.

If this seems a great deal of worry to expend over mere subspecies, consider that there is growing evidence that North America's red crossbill is actually seven or eight separate species, confusingly similar but ecologically (and reproductively) distinct. Recent research suggests that each has evolved to exploit a particular conifer or group of conifers—trees with reliable, annual cone crops that are (at least in the absence of introduced competitors like squirrels) unavailable to other animals. The various crossbills provide an unusually clear window into the workings of natural selection and the constant jockeying that goes on among organisms trying to gain a slight advantage in the difficult task of living.[4]

If there is a symbol that the Appalachians of Newfoundland are a world apart, it must be the yellow diamond road signs one sees along Route 430, the main north-south avenue that dances between the coast and the mountains. They show a silhouette, not of the deer of the Smokies or the moose of Maine, but of a caribou, antlers branching manically above a stylized head.

Only in Newfoundland are caribou a common Appalachian mammal. A relict herd, it is true, survives on Mont Albert in the Shickshocks of Quebec's Gaspé, and a few were returned briefly to Maine in a failed reintroduction, but the Long Range Mountains hold thousands.

These are woodland caribou, one of four subspecies in North America. Somewhat larger than the more familiar "barren-grounds" caribou of Alaska and Arctic Canada, they are creatures of the taiga, the stunted, boreal forest of spruce and aspen that covers the midsection of Canada (they were once also found as far south as northern New England and the northern Great Lakes, but hunting, timbering and deer-carried brainworm eliminated them there).

When I went to the Long Range Mountains for the first time, I'd hoped to find caribou; the thought of seeing such a visible link to the Pleistocene in the Appalachians had a powerful lure. But when I asked in Gros Morne what backcountry areas would be best, the ranger just smiled. The caribou, she said, were easy to see in winter, when they came down from the mountains and were often near the road. In summer, though, they disappear back into the fastness of the Long Range. "We don't really know where they go," she said, apologetically, and suggested I come back in December, to mix cross-country skiing with my caribou-watching.

I thanked her, but I didn't tell her she'd given me a greater gift than any caribou sighting. When I hiked the mountains above Gros Morne's soaring fjords in the days that followed, I was almost glad that the horizons were empty of antlered heads. The knowledge that there is still one part of the Appalachians so wild and unsettled that herds of caribou can simply vanish into them, for months on end, is beyond price.

North American caribou and Eurasian reindeer are simply different subspecies of the same animal, but the reindeer has a long history of domestication in Scandinavia, while the caribou does not; besides providing meat and skins, reindeer are a source of milk and harnessed labor for their herders. Struck by the possibilities they offered, Dr. Wilfred T. Grenfell imported a small herd of reindeer to northern Newfoundland in 1908.

Grenfell was—and still is—an almost legendary figure in the small outports and villages of the Northern Peninsula. English by birth, he turned to medical missionary work in 1887 while still a student, spending five years as surgeon on a hospital ship in the North Sea. Grenfell came to Labrador and Newfoundland in 1892, when this was an incredibly isolated and desolate region, wracked by poverty, malnutrition and disease. Grenfell, and the International Grenfell Association he created, established hospitals (including the large, modern facility in St. Anthony) and hospital ships, nursing stations, orphanages, schools and even a lumber mill. The doctor himself traveled tirelessly, often under hazardous conditions; once, trapped on an ice floe, he was able to stay alive only by eating several of his beloved sled dogs.

Grenfell was also constantly looking for ways to provide alternate sources of income for the local people, and he thought that reindeer would be an excellent supply of meat and hides. (Native caribou had always been hunted by Newfoundlanders, of course, but by the beginning of the twentieth century their numbers were in serious decline, and they couldn't be treated as livestock, like the tractable reindeer could.) The 250 reindeer that Grenfell imported arrived complete with Lap (Sami) herders and quickly increased, so that by 1914 there were more than two thousand.

After that, however, things began to fall apart. It may be that the residents of the Northern Peninsula never quite thought of reindeer as stock instead of meat-on-the-hoof; older folks I talked to, who remember the reindeer herds (or who remember their parents talking about them) told me that many were poached, and others were killed by sled dogs. In any event, Dr. Grenfell's dream of reindeer ranching died out. Another scheme, this one by a timber company to maintain reindeer herds near logging camps as a fresh meat supply, likewise failed—but only after the reindeer were herded south from St. Anthony and across the Long Range, spreading an Old World parasite among native caribou in their wake.

Dr. Grenfell's heart was in the right place, but perhaps it's just as well that the reindeer project faltered. The windswept barrens and high hills of the Long Range are an untamed and rugged place, and I think it fitting that only wild caribou make their living there. Domestication sits poorly on a land such as this.

I followed one of the ghostly old caribou trails, a hundred years vacant, up into the hills of Belle Isle as evening came on. The sun was engulfed in the fog bank waiting just offshore, casting everything in a diffuse orange light. I sat on the highest hill, rocky and scraped clean of soil, the brittle lichen crackling and snapping underfoot, and looked down over the Appalachians' finale.

The wind had dropped to stillness, and the silence was more profound than anything I'd ever experienced; even the hush of the waves was muted. The absence of sound seemed somehow to make my vision abnormally clear. I could see the stream a mile or more away where the irises and roseroot grew in thick masses, and off to the east, near the edge of the fog, the last wheeling flocks of kittiwakes, coming in to the cliffs to roost. In the thick golden light, the ponds that rimmed the hill shone like scattered suns.

Dusk settled in, a blue hour wrapped in mist, and far off the yellow beam of the lighthouse began its march, around and around, slicing the gloom. A savannah sparrow, hidden in the folds of the hill near me, whispered its single, reedy call note in almost perfect time with the light's revolutions: *Seep. Seep. Seep.* Then I heard the rattle of its wings as it flew to roost, and the night was even more arrestingly silent than it had been before. With a sigh I rose and turned my footsteps south.[5]

Bunchberry and Wood Sorrel

Epilogue

MOUNT ABRAHAM, VERMONT

Summer in the Green Mountains is a sweet season, but it does not linger. The hobblebush leaves were already turning bronze, even though August was only half-finished, and here and there a red maple was getting a jump on autumn, teasing the forest with sparks of orange. The trail from Lincoln Gap to the top of Mount Abraham, more than four thousand feet up, was lined with bunchberry plants, resplendent with their clusters of bright crimson fruit.

There were five of us—me, my old friend Rick, his two young daughters and their aging yellow Lab. This was the girls' first real mountain hike, at least under their own steam; their parents had carried them up peaks when they were younger, but they'd never climbed one on their own. Krista, who was tugging me up the trail by one hand, was nearly seven; Samantha, who had hold of the other, was almost five.

It was a muggy, somewhat hazy day, with a promise of rain if the clouds ever got their act together. The girls were eager, always pushing ahead, heedless of puddles and scratched knees, finding the sorts of things kids always find in the woods—odd mushrooms and pretty rocks, deer tracks in the mud, the wings of a dragonfly caught in a spider's web. We taught them how to identify wood sorrel and let chew the tart, citric leaves; thereafter, Samantha grazed her way to the top, stopping every few yards to nibble.

The trail rose through northern hardwood forest, my favorite Appalachian plant community, with its graceful birches and weathered maples shimmering in the light breeze, and the ground cover of clintonia and bunchberry. The paper birches in particular were lovely, the fresh bark tinged with a shade of pink and the old, peeling bark a chalky gray-white. Then we passed up into balsam fir and red spruce, and the children laughed with delight as the trees shrank smaller and smaller the higher we climbed, until at the peak, even tiny Samantha was taller than the forest.

The summit of a mountain is always a revelation; you can never believe the world is quite so big, or even half so beautiful. We were surrounded by hills of blue, standing like cutouts from the valleys and ranged in disappearing ranks in all directions. We stood, it seemed, at the center of a universe of mountains and felt that the cosmos was wheeling around us.

Overhead the clouds deepened, and the first drops fell as we turned downhill. Not long after, the thunder began, and we hurried as much as the muddy trail allowed.

Then the thunder roared again, much closer this time, and we could not help but stop to listen. It was wonderfully deep and intricately resonant, layers of sound like the layers of hills around us. It boomed down into the valleys and echoed back up to the heights, taking character and depth from the land over which it rang: now quiet and gentle as the mist, now ripping like torn canvas so that we jumped in spite of ourselves.

The thunder rolled through the gaps and over the peaks, making our bones rumble in concert, sharpening its teeth on the rocky cliffs, softening itself on the forested slopes, uniting alpine tundra with valley cornfield, air with rock, and the immensity of time with the rhythm of our pulse—a voice that spoke for the mountains of my heart, a song to span the centuries and the miles.

Notes

Chapter 1. The Supple Rock

1. While the fact of plate movement has been repeatedly demonstrated, debate remains on the mechanism, the engine that makes it go. One school of thought holds that convection currents, like the roiling of a pot of hot water, make cells within the mantle churn. Others believe that the cooler, denser margins of the oceanic plates subduct beneath the continental plates largely because of their own weight.

2. An alternate theory holds that rather than Africa, an island terrane— caught in the squeeze between the two colliding plates—actually hit the eastern edge of the North American Plate and crumpled to form the Appalachians, much as happened more recently along the Pacific coast.

Chapter 2. Islands in the Sky, Travelers on the Wind

1. A scientific (or "Latin") name has two main parts. The first, the genus name, is shared by other closely related organisms; all North American falcons, for example, are members of the genus *Falco*. The second part is the species' specific name, which is always lowercase; in the case of the falcon known as the merlin, *columbarius*.

   Finally, if populations within a species are sufficiently different from one another, scientists may further divide them into subspecies. North American merlins are split into three subspecies largely on the basis of color; the form found in the Appalachians is *Falco columbarius columbarius*, while the pale prairie subspecies to the west is *F.c. richardsonii* (subspecific names are often abbreviated this way or referred to by their subspecific name only).

   Scientific names may seem like an annoying affectation, but they serve a vital purpose by making identification precise and accurate. Unlike common names that may vary with region or country, each type of living thing is assigned only one scientific name, permitting specialists from different regions or languages to communicate without confusion.

2. *Couchant* is a term from medieval heraldry referring to an animal resting on its stomach, forelegs stretched out in front.

3. Of the more than sixty species of true alpine plants in the White Mountains, almost 70 percent are also found in the Arctic. This is understandable, since tundra once grew far south of New England. What is more intriguing is the presence of a number of otherwise Rocky Mountain or Sierra Nevadan plants in the northern Appalachian peaks.

These "mountain disjuncts," as botanists call them, probably once grew straight across the continent in the narrow tundra band below the glaciers. When the ice sheets melted, some plants moved north and reoccupied the Arctic; they include the small fireweed known as river-beauty, which is found from Newfoundland to Alaska and south through the Rockies. Not all the tundra species could tolerate conditions in the Arctic, however, but did manage to survive at both ends of the band—the Rockies in the West and the New England peaks, the Shickshocks and the Long Range Mountains in the East. They include *Diapensia* and the low-growing, yellow wildflower called mountain-avens.

Chapter 3. The Wooded Sea

1. Sadly, those ancient hemlocks have since fallen victims to hemlock woolly adelgids, and were removed by the U.S. Forest Service.

2. According to calculations by Audubon, a flock of roughly a billion passenger pigeons would eat 8.7 million bushels of mast each day.

3. This reliance of migrant songbirds on autumn fruit is a worrisome factor in the already tragic story of the flowering dogwood. Among the most common of understory trees in the central and southern Appalachians, flowering dogwood suffered unprecedented losses to a fungus known as dogwood anthracnose. Some areas have lost more than half their dogwoods; others, virtually all.

Only discovered in the mid-1970s and not isolated by scientists until 1991, anthracnose spread throughout the Appalachians with astonishing speed. In just a short time it killed millions of dogwoods, particularly those growing at higher altitudes in shady, damp surroundings—in other words, the mountains stands for which the southern Appalachians have long been famous. (Dogwoods in yards and along road edges, which are exposed to more sun and higher temperatures, are disfigured by the disease but usually don't die.) Acid precipitation and ground-level air pollution may be aiding the fungus by weakening the dogwood's defenses.

While experimental, resistant dogwoods have been replanted in some hard-hit areas like the Catoctin Mountains, foresters are bracing themselves for the functional loss of the dogwood—and wondering what that loss will mean for migrating songbirds that can't find their usual fat supply.

Chapter 4. From Fertile Waters

1. Three other salmonids are native to the northern Appalachians: lake trout; both landlocked and sea-run populations of Atlantic salmon from New York north; and Arctic char in northern New England and Atlantic Canada. The last includes several unique landlocked populations, among them the small "blueback trout" of Maine's Rangely Lakes region, and the "Sunapee trout" of Sunapee Lake, New Hampshire, and a handful of other New England lakes.

2. In birds such breeding aggregations are known as leks. Studies of lekking species like some grouse and shorebirds show that dominant males tend to occupy the central (and presumably best) territories, which are sought out by females that must run a gauntlet of subdominant males in peripheral territories in order to reach them. Whether this is true for a peeper marsh with hundreds or thousands of frogs remains to be seen.

Chapter 5. Keeping Faith with the North

1. The adelgid is not the first time the eastern hemlock has suffered an enormous blow. Studies of pollen from lake-bottom cores and other fossils indicate that about forty-eight hundred years ago, hemlocks all but vanished across their entire range; while the tree rebounded somewhat over the next thousand years, it remains less abundant today than in the past.

   One possible cause may have been the two species of caterpillars known as hemlock loopers, which can defoliate hemlocks and cause localized mortality. Perhaps the loopers did not originally exist within the hemlock's range, in which case the trees would have had no defense against them when they arrived. Fungal outbreaks are another possibility. In any event, the fact that hemlocks survived one catastrophic collapse gives hope that they may survive the adelgid—even if recovery takes a millennium.

2. You can distinguish the two by skull characteristics, if you're so inclined. The anterior supraorbital process—a bone protrusion on the leading edge of the cheekbone—is pronounced in the eastern cottontail but miss-

ing in the Appalachian. Also, the suture between the nasal bones the rest of the skull is jagged in Appalachians but smooth in easterns.

Chapter 6. Thunder, Dimly Heard

1. The debate over the relationship between elk and red deer continues, with more recent genetic evidence supporting the elk's previous classification as a distinct species.

2. The Allegheny woodrat, *Neotoma floridana magister*, was until the 1950s considered a distinct species from the eastern woodrat of the Southeast and southern plains. Then, because of similarities in external appearance and skull characteristics, the two were lumped together by mammalogists. More recent DNA studies, however, suggest the Appalachian population is, as originally thought, a separate species.

Chapter 7. Roots in the Hills

1. There is at least a possibility that even the Norse were not the first Europeans to see Newfoundland. Irish tradition holds that the fifth-century monk Saint Brendan, along with seventeen followers, set out in an ox-hide curragh to seek the Promised Land of the Saints, said to lie beyond the western horizon. The medieval text *Navigato Sancti Brendani Abbatis* tells of their seven-year voyage to such wonders as the Paradise of Birds, populated by fallen angels that God had changed to Latin-speaking fowl, the Island of Grapes and the Island of Delights. They constantly bump into other monks and saints (one of whom has been living on his rock for thirty years with only spring water to sustain him), and even find Judas Iscariot serving penance on a lonely rock. At last they come to the Promised Land of the Saints, returning with fruit and jewels.

Obviously, the Voyage of Saint Brendan cannot be taken as a historical document, even within the liberal bounds accorded the Viking sagas. But some parts of the account ring true—the sailors trapped in a "coagulated sea" that may have been pack ice, seeing a volcano on what could have been Iceland, or finding a huge mountain the color of a silver veil but hard as marble, with a hole in it big enough to sail through. Fortunately they did not, for the top collapsed moments later. It is an uncommonly poetic description of a decaying iceberg.

In 1976–1977 Irish explorer Tim Severin and a crew of four crossed the north Atlantic in a thirty-six-foot leather curragh built to medieval specifications. They island-hopped in much the same way the Norse did, from Ireland to the Faroes, then to Iceland, around the cape of Greenland and finally to Newfoundland. Of course, this proves only that men could make the crossing, not that Saint Brendan did. While many scholars accept that Brendan did make one or more sea voyages, many believe he is likely to have traveled to the Azores instead of North America.

Chapter 8. Pinus and Castanea

1. By way of one small and revealing example, consider three formerly widespread insects—the chestnut case-bearer moth, the Pennsylvania chestnut yponomeutid moth and the American chestnut clearwing. The larvae of the first two fed on chestnut foliage, while the clearwing larva bored into damaged trunk tissues, and the blight robbed them of their universe. The case-bearer and the yponomeutid haven't been seen in decades (a century, for the latter), and are likely extinct, but the clearwing was rediscovered in 1985. It is, ironically, now a pest in plantations of backcrossed American chestnuts being field-tested for restoration.

2. Some of these extralimital stands were remarkable, and provided the only surviving hint of what the Appalachian chestnut forests must have looked like. One in West Salem, Wisconsin, numbered about six thousand chestnuts, including a large number of canopy trees up to two feet in diameter and more than a century old. The West Salem stand dates to the 1880s, when farmer Martin Hicks planted chestnut seeds his mother brought from central Pennsylvania. Other extralimital chestnut groves could be found in Michigan, Maine and Oregon.

   Although the blight missed most of these trees at first, more recently it has appeared in many of the groves. Most of the mature trees in the West Salem stand, for example, have been killed by the blight, despite heroic efforts by plant pathologists to treat them with hypovirulent strains. More isolated trees have survived, by their very isolation; in 2012 a ninety-five-foot-tall chestnut was discovered in Maine, the tallest east of the Mississippi, and one of several hundred scattered chestnuts known to grow in the state.

   The value of these chestnut stands goes beyond aesthetics, of course. They provide the only way to study American chestnut ecology, including how the tree interacts with other hardwoods. Scientists working in West Salem, for example, where soil and climate conditions resemble those in many areas of the Appalachians, have found that chestnuts are aggressive, displacing oaks and hickories that were already established on the site. While few of the seeds sprout in undisturbed woodland, saplings can survive for long periods of time as low understory plants, waiting for a gap in the canopy to open. When that happens, they experience rapid growth, shooting up and monopolizing the light—a strategy also followed by many rain forest trees. A chestnut tree sixty feet tall and a foot thick may be just twenty-five years old, a fifth the age of a similarly sized oak.

3. However hopeful the chestnut's story, these are perilous days for many other Appalachian trees, and some pests and potential pathogens—should they appear in the mountains—could put the chestnut blight to shame for raw destruction.

We've already seen the damage wrought by the hemlock woolly adelgid. Even worse, for its speed and virulence, has been the emerald ash borer, an Asian beetle first detected near Detroit in 2002. Having virtually annihilated native ash trees in Michigan, it has spread throughout the Midwest and East and has been found across essentially the entire Appalachian region.

The emerald ash borer kills only ashes, however. The Asian long-horned beetle, another hitchhiker brought on untreated wood from China, eats an utterly catholic diet of hundreds of hardwoods and conifers, which die from the dime-sized tunnels of its wood-boring larvae. The only way to control it is to cut down, pulp and burn every tree within the infected zone. That's possible in tree-lined urban neighborhoods, some of which have been completely denuded in the effort, but should this beetle establish itself in the mountain forests (as some foresters assume it has) the result would be devastating.

Then there are diseases. Butternut canker, an Asian pathogen, has reached almost every stand of the tall "white walnut," as it is sometimes known, killing most of the infected trees within ten or fifteen years. Few people have heard of butternut or can recognize it in the woods, so it's received little of the attention given chestnuts, although there is a last-ditch effort to find resistant trees, and to graft twigs from them to black walnut root stock to propagate that gene line.

Oak forests have been suffering from several mysterious maladies. "Oak decline" is the name given to a complex of problems—involving insects, fungi and environmental conditions—that cause widespread death not only in oaks but also in birches, ashes, maples and beech. Oak wilt is a fungal disease, spreading ever more widely in the East, that is utterly nondiscriminating, killing all species of oaks, usually within a year or two of infection.

Yet bad as that may be, the real nightmare bug is sudden oak death, a fungal disease that appeared in California and Oregon and has killed millions of western live oak species. Lab tests show that eastern species, especially those in the red oak clan, are exceptionally susceptible—and because the fungus, *Phytophthora ramorum*, also infects ornamental shrubs like rhododendron and mountain laurel, which serve as hosts, it has easy passage via nursery stock sold nationwide. Quarantine efforts have been leaky at best; stock from one infected nursery was shipped throughout the southern Appalachians in 2003–2004, and the fungal spores have been detected in waterways, suggesting it is present, though not yet causing an epidemic.

In the view of many experts, *Phytophthora* represents the single greatest threat to the forests of the East, especially the biodiverse hardwood stands of the southern Appalachians, where red oaks may comprise 80 percent of the canopy. It's already out there, and no one knows if some environmental condition is preventing it from spreading, or if it simply hasn't yet reached an infectious tipping point.

4. There had been other stands of old white pines, but the powerful hurricane of September 1938 flattened many of them—trees that stood so much taller than their neighbors that they had little shelter from the blast. The Harvard Forest near Winchester, New Hampshire, was one casualty, and perhaps the most grievous loss. The best example of old-growth mixed forest in New England, it contained some pines and hemlocks more than three hundred years old, of which a small remnant remains. In the summer of 1989, three tornadoes toppled most of the old-growth trees in Cathedral Pines, which is now a study site for forest recovery and restoration.

Chapter 9. Winter in the High Lands

1. This surface-to-volume ratio is a stumbling block for any warm-blooded animal. Since surface (length times length) increases more slowly than volume (length times length times length), bigger animals will always have more heat-generating mass than smaller ones. This is probably why the larger individuals of many species are found in the colder regions of their range, a biological principle known as Bergmann's Rule. From a heat conservation perspective, in other words, it is easier to be a deer than a deer mouse.

2. I once came across one of these fecal plugs, which are expelled by the bears shortly after leaving their dens in spring. It was nearly eight inches long (some may measure a foot in length) and was composed mostly of twigs as thick as cooked spaghetti, large numbers of white pine needles

and was bound together with blackish intestinal secretions. Considering where it came from, it was the most uncomfortable looking thing I've ever seen, and I try not to think about it too much.

3. Wood frogs and spring peepers, both common amphibians in the northern Appalachians, show this same remarkable tolerance to freezing. Unlike many frogs, which dig into lake sediments or burrow beneath the frost line, they winter just beneath the surface of the soil, where they are subjected to prolonged, intense freezing. They survive by larding their cells with astounding concentrations of a form of sugar known as glycogen, which is stored within the liver between bouts of freezing. The gray treefrog, another freeze-resistant species found as far north as southern Maine, can survive temperatures as low as minus twenty degrees Fahrenheit thanks to this defense.

4. Interestingly, by the late 1980s the widespread, roughly biennial invasions of evening grosbeaks into the East essentially ended, and no one is quite sure why. The best explanation is that, while the adult grosbeaks feed on seeds, their chicks depend on insects—especially spruce budworm, a cyclical pest of Canadian conifer forests. The collapse in evening grosbeak invasions coincided with the decline of budworm outbreaks in the 1980s (and declining populations of so-called budworm specialists like blackpoll, Cape May and bay-breasted warblers). By the 2010s, it appeared the budworm cycle was on an upward trajectory, and reports of evening grosbeaks in the Northeast were again on the rise, along with increasing numbers of the budworm-dependent warblers.

5. While feeding domestic mice to northern owls, luring them close for photographs or observation, used to be a common practice, it's frowned on these days, and I certainly wouldn't do it again as I did in this case, in the early 1990s. Already dangerously naive about humans, the owls can quickly become habituated by feeding, placing them at further risk.

6. Litter size varies widely in the Appalachians. Pennsylvania bears, as noted in Chapter 6, have unusually large litters. Farther north, where food is less abundant and average adult weights are considerably less, the average number of cubs declines as well. Taken as a whole population, most Appalachian black bears give birth to two cubs every two years.

Chapter 10. Seeing a Forest for the Trees

1. As an illustration that nothing in nature is quite as simple as it appears, consider this: Some of the chemicals manufactured by oak trees as a de-

fense against insect attacks appear to actually benefit gypsy moths. Tannins, which are common in oak foliage, stunt caterpillar growth and interfere with their reproduction, but they also damage the ability of the wilt virus to "recognize" the gypsy moth as its correct host. The higher the tannin content of the leaves on which it feeds, the greater the caterpillar's resistance to wilt disease. Those feeding on oaks do not grow as large as those on tannin-poor leaves like aspen, but they are as much as twenty times less likely to die of the virus.

2. New York furbearer biologist Ben Tullar, Jr., who believes that coyotes are a native Northeastern species, argues that most or all of the nearly one hundred "wolves" turned in for bounty in New York between 1871 and 1897 were eastern coyotes, as were wolf-like canids shot between 1906 and 1920.

3. Concern about neotropical migrants has also forced a rethinking of timbering practices. Huge clear-cuts, which have fallen into such disfavor that Maine banned them in 1991, may in fact help deep-woods songbirds, since they eventually regenerate into solid tracts of forest with proportionally less edge; this reduces the impact of nest predators and cowbirds. By restricting clear-cuts to 250 acres or less, a few biologists claim, the Maine law may actually foster fragmentation and worsen the situation for songbirds.

Chapter 11. Ultima Thule

1. The name may also come from a corruption of the French *baie qu' appelle*, "What's this berry called?"

2. In 2012, a coyote hunter on Newfoundland's Bonavista Peninsula shot what proved to be an eighty-two-pound wolf, which wildlife authorities believe recolonized the island on winter ice from Labrador. It was the first wolf confirmed in Newfoundland since the 1930s. Wolves have also recently returned to New Brunswick for the first time since 1876, crossing the St. Lawrence River on winter ice from the north shore in Quebec.

3. Newfoundland's mammalian fauna is unusual in another respect—the high number of carnivores compared to herbivores. Seven of the fourteen resident native mammals were carnivores (gray wolf, lynx, black bear, red fox, short-tailed weasel, river otter and pine marten), while only five (beaver, muskrat, meadow vole, Arctic hare and caribou) are herbivores. The remaining two, both bats, are insectivores.

It is theorized that this skewed ratio in favor of meat-eaters stems from Newfoundland's island status—carnivores tend to be bigger, wider-ranging and more active than most herbivores, and thus most likely to reach an island by water or icepack. Compare this predator-herbivore ratio with that of Pennsylvania, where in presettlement days there were roughly seventeen carnivore species, twenty-seven herbivores and twenty-one insectivores (bats, shrews and moles).

4. The contention that a "Northeastern" race was driven to extinction by logging, which was based on an examination of museum specimens, is now largely discounted by ornithologists—but a previously overlooked resident Appalachian form, with a distinct call type, was finally confirmed in 2011. Its core range is from southern New York to northern Georgia, overlapping the highland range of red spruce, but it irrupts as far north and west as Nova Scotia and the Pacific Northwest.

The Newfoundland form of the red crossbill, meanwhile, has continued to decline in the face of red squirrel competition, and is now listed as endangered by Canada.

5. One more change came to Belle Isle in 2001, when the light station was destaffed and fully automated.

Bibliography

Alerstam, Thomas. *Bird Migration*. New York: Cambridge University Press, 1990.

Allen, J. A. *The American Bisons Living and Extinct*. 1876. Reprint, Salem, N.H.: Arno Press, 1974.

Anderson, Cindy. "The Fight Against Blight," *Nature Conservancy* (November/December 1992), 8–9.

Baird, Spencer Fullerton. *Mammals of North America: The Description of Species Based Chiefly on the Collections in the Museum of the Smithsonian Institution*. Philadelphia: J. B. Lippincott, 1859.

Banks, Richard C. "The Decline and Fall of the Eskimo Curlew, or Why Did the Curlew Go Extaille?" *American Birds*, Vol. 31 (March 1977), 127–34.

Barbour, Michael G., and William D. Billings, eds. *North American Terrestrial Vegetation*. New York: Cambridge University Press, 1988.

Bendick, Robert. "Zebra Mussel Management," *New York Conservationist*, Vol. 47 (July/August 1992), 42–43.

Benkman, Craig W. "A Crossbill's Twist of Fate," *Natural History*, Vol. 101 (December 1992), 38–44.

———. "The Evolution, Ecology and Decline of the Red Crossbill in Newfoundland," *American Birds*, Vol. 47 (Summer 1993), 225–29.

Bent, Arthur Cleveland. *Life Histories of North American Birds of Prey*, Part II. Washington, D.C.: U.S. Printing Office/Smithsonian Institution Bulletin 170, 1938.

———. *Life Histories of North American Jays, Crows and Titmice*, Part I. Washington, D.C.: U.S. Printing Office/Smithsonian Institution Bulletin 191, 1946.

Blankenship, Karl. "Our Disappearing Songbirds," *Apprise*, Vol. 13 (September 1993), 42–47.

Blankinship, David R., and Kirke A. King. "A Probable Sighting of 23 Eskimo Curlews in Texas," *American Birds*, Vol. 38 (November/December 1984), 1066–67.

Blockstein, David E., and Harrison B. Tordoff. "Gone Forever: A Contemporary Look at the Extinction of the Passenger Pigeon," *American Birds*, Vol. 39 (Winter 1985), 845–51.

Boatwright, Vicki, and Karen Ballentine. "Red Wolves Return to Great Smoky Mountains," *Fish & Wildlife News*, United States Fish & Wildlife Service (January/February 1990).

Brain, Jeffery P. "From the Words of the Living: The Indian Speaks," in *Clues to America's Past*. Washington, D.C.: National Geographic Society, 1976.

Brauning, Daniel W., ed. *Atlas of Breeding Birds in Pennsylvania*. Pittsburgh: University of Pittsburgh Press, 1992.

Brooks, Maurice. *The Appalachians*. Boston: Houghton Mifflin Co., 1965.

Burks, Ned, and Chris Fordney. "Battle for the Blue Ridge," *Washington Post Magazine*, October 31, 1993, 14–20, 29–31.

Burns, D. A., J. A. Lynch, B. J. Cosby, M. E. Fenn, and J. S. Baron, U.S. EPA Clean Air Markets Division. *National Acid Precipitation Assessment Program Report to Congress 2011: An Integrated Assessment.* Washington, D.C.: National Science and Technology Council, 2011.

Burroughs, John. *Signs and Seasons.* 1886. Reprint, New York: Harper Colophon, 1981.

Burzynski, Michael, and Ann Marceau. "Plants of the Serpentine Barrens," *Wildflower* (Winter 1993), 38–43.

Campbell, Carlos C., William F. Hutton, and Aaron J. Sharp. *Great Smoky Mountains Wildflowers.* 4th ed. Knoxville: University of Tennessee Press, 1977.

Catlin, David T. *A Naturalists' Blue Ridge Parkway.* Knoxville: University of Tennessee Press, 1984.

Caudill, Harry M. *Night Comes to the Cumberlands.* Boston: Atlantic Monthly Press/Little, Brown, 1962.

Christine, Mariette D. "American Chestnut: Down But Not Out," *Popular Science,* Vol. 243 (July 1993), 75.

Cochran, M. Ford. "Chestnuts: Back From the Brink," *National Geographic,* Vol. 177 (February 1990), 128–40.

"Confounding the Rodents," *Discover,* Vol. 13 (November 1992), 20.

Connelly, Thomas L. *Discovering the Appalachians.* Harrisburg, PA: Stackpole Books, 1968.

Conners, John A. *Shenandoah National Park: An Interpretive Guide.* Blacksburg, VA: McDonald & Woodward Publishing Co., 1988.

"Conserving a Coyote in Wolf's Clothing?" *Science News,* Vol. 139 (June 15, 1991), 374–75.

Coolidge, Philip T. *History of the Maine Woods.* Bangor, ME: Furbush-Roberts Printing, 1963.

Cowen, Ron. "Brooding over Australian Frogs," *Science News,* Vol. 137 (March 3, 1990).

Daniels, G. G. "A Possible Sight Record of Eskimo Curlews on Martha's Vineyard, Mass.," *American Birds,* Vol. 26 (October 1972), 907–8.

Dann, Kevin T. *Traces on the Appalachians: A Natural History of Serpentine in Eastern North America.* New Brunswick, NJ: Rutgers University Press, 1988.

De Bleiu, Jan. *Meant to Be Wild.* Golden, CO: Fulcrum Publishing, 1991.

Delcourt, Paul A., and Hazel R. Delcourt. *Long-term Forest Dynamics of the Temperate Zone.* New York: Springer-Verlag, 1987.

Dickerman, Robert W. "The 'Old Northeast' Subspecies of Red Crossbill," *American Birds,* Vol. 41 (Summer 1987), 188–94.

Dixon, E. James. *Quest for the Origins of the First Americans.* Albuquerque: University of New Mexico Press, 1993.

Doolittle, Jerome. *The Southern Appalachians.* New York: Time-Life Books, 1975.

Doutt, J. Kenneth, Carolina A. Heppenstall, and John E. Guilday. *Mammals of Pennsylvania*. Harrisburg, PA: Pennsylvania Game Commission, 1966.

Durant, Mary, and Michael Harwood. *On the Road with John James Audubon.* New York: Dodd, Mead & Co., 1980.

Faanes, Craig A., and Stanley E. Senner. "Status and Conservation of the Eskimo Curlew," *American Birds*, Vol. 45 (Spring 1989), 257–59.

Fenton, M. Brock. *Bats.* New York: Facts on File, 1992.

Fowells, H. A. *Sylvics of Forest Trees of the United States.* Washington, D.C.: United States Department of Agriculture/Agriculture Handbook 271, 1965.

Frome, Michael. *Strangers in High Places: The Story of the Great Smoky Mountains.* New York: Doubleday, 1966.

Fulbright, Dennis. "Treatment of the West Salem Stand with Hypovirulence," *The Bark*, newsletter of American Chestnut Foundation (May–June 1993), 1–3.

Genoways, Hugh H., and Fred J. Brenner, eds. *Species of Special Concern in Pennsylvania.* Pittsburgh: Carnegie Museum of Natural History, 1985.

Geology, Topography, Vegetation: Gros Marne National Park, GSC Miscellaneous Report 54. Ottawa: Geologic Survey of Canada, 1992.

Gilbert, Bil. "Coyotes Adapted to Us, Now We Have to Adapt to Them," *Smithsonian* (February 1991), 65–79.

Gill, Frank B. *Ornithology.* New York: W. H. Freeman & Co., 1990.

———."Whither Two Warblers?" *Living Bird Quarterly*, Vol. 4 (Autumn 1983), 4–7.

Gollop, J. Bernard. "The Eskimo Curlew," in *Audubon Wildlife Report 1988-1989.* San Diego: Academic Press, 1988.

Gould, Stephen Jay. "Abolish the Recent," *Natural History*, Vol. 100 (May 1991), 16-21.

Graham, Frank, Jr. "2001: Birds That Won't Be With Us," *American Birds*, Vol. 44 (Winter 1990), 1074–81, 1194–99.

Griffin, John W. *Investigations in Russell Cave.* Washington, D.C.: National Park Service, 1974.

Grinnel, George Bird. *Last of the Buffalo.* New York: Scribner's Magazine, 1892.

Griscom, Ludlow, and Alexander Sprunt, Jr., eds. *The Warblers of North America.* New York: Devin-Adair Co., 1957.

Hagar, Joseph A., and Kathleen Anderson. "A Sight Record of Eskimo Curlew (*Numenius borealis*) on West Coast of James Bay, Canada," *American Birds*, Vol. 31 (March 1977), 135–36.

Hagen, John M. III, and David W. Johnston, eds. *Ecology and Conservation of Neotropical Migrant Landbirds.* Washington, D.C.: Smithsonian Institution Press, 1992.

Halfpenny, James C., and Roy Douglas Ozanne. *Winter: An Ecological Handbook.* Boulder, CO: Johnson Publishing Co., 1989.

Hall, George A. *West Virginia Birds.* Pittsburgh: Carnegie Museum of Natural History, 1983.

———. "Population Decline of Neotropical Migrants in an Appalachian Forest," *American Birds*, Vol. 38 (January/February 1984), 14–18.

Hall, R. E., and K. R. Kelson. *Mammals of North America*, Vol. II. New York: Ronald Press Co., 1959.

Hamilton, William J., Jr., and J. O. Whitaker, Jr. *Mammals of the Eastern United States*. 2nd ed. Ithaca, NY: Cornell University Press, 1979.

Harrison, Hal H. *Wood Warbler's World*. New York: Simon & Schuster, 1984.

Hayden, Arnold H. "The Eastern Coyote Revisited," *Pennsylvania Game News*, Vol. 60 (December 1989), 12–15.

Hayes, John P. "How Many Species of Woodrats?" *New York Conservationist* (March/April 1989), 39.

Hayman, Peter, John Marchant, and Tony Prater. *Shorebirds: An Identification Guide*. Boston: Houghton Mifflin Co., 1986.

Heinrich, Bernd. "A Birdbrain Nevermore," *Natural History*, Vol. 102 (October 1993), 50–57.

Hendricks, Walter. "Where a Heavy Body Is Likely to Sink," *Audubon*, Vol. 83 (September 1981), 98–111.

Herrington, John. "The American Chestnut Foundation," *Environment*, Vol. 34 (December 1992), 4–5.

Hicks, Alan. "Whatever Happened to the Allegheny Woodrat?" *New York Conservationist* (March/April 1989), 34–38.

Holloway, Marguerite. "Musseling In," *Scientific American*, Vol. 267 (October 1992), 22–23.

Hoover, Jeff. Nesting Success of Wood Thrush in Fragmented Forest. Unpublished thesis, 1992.

Houck, Rose. *A Natural History Guide: Great Smoky Mountains National Park*. Boston: Houghton Mifflin Co., 1993.

Hutto, Richard L. "Is Tropical Deforestation Responsible for the Reported Declines in Neotropical Migrant Populations?" *American Birds*, Vol. 42 (Fall 1988), 375–79.

Jackson, Laura E. *Mountain Treasures at Risk: The Future of the Southern Appalachian National Forests*. Washington, D.C.: The Wilderness Society, 1989.

Jennings, Francis. *The Ambiguous Iroquois Empire*. New York: W. W. Norton, 1984.

Karason, William H. "In the Belly of the Bird," *Natural History*, Vol. 102 (November 1993), 32–37.

Kephart, Horace C. *Our Southern Highlanders*. 1913. Reprint, Knoxville: University of Tennessee Press, 1984.

Kilmer, Joyce. *Trees and Other Poems*. New York: Doubleday, 1914.

Komhisen, Laurel. "The Working Forests of Maine," *Living Bird*, Vol. 12 (Autumn 1993), 14–19.

Krohn, William B., and Kenneth D. Elowe. "Do the Pieces Fit? Understanding a Harvested Fisher Population," *Maine Fish and Wildlife*, Vol. 35 (Fall 1993), 6–11.

Kulick, Stephen, P. Salmansohn, M. Schmidt, and H. Welch. *The Audubon Society Field Guide to the Natural Places of the Northeast: Inland.* New York: Pantheon Books, 1984.

Kuznick, Frank. "What Difference Does the Dogwood Make?" *National Wildlife,* Vol. 31 (April/May 1993), 46–50.

———."American's Aching Mussels," *National Wildlife,* Vol. 31 (October/November 1993), 34–39.

"Last Stand: The History and Ecology of the American Chestnut in West Salem, WI." *The Bark,* newsletter of American Chestnut Foundation (March–April 1993), 1–4.

Lawren, Bill. "Singing the Blues for Songbirds," *National Wildlife,* Vol. 30 (August/September 1992), 4–11.

Lawrence, Susannah, and Barbara Gross. *The Audubon Society Field Guide to the Natural Places of the Mid-Atlantic States: Inland.* New York: Pantheon Books, 1984.

Laycock, George. "Fungus vs. Fungus," *Audubon,* Vol. 86 (May 1984), 42–45.

Leopold, Aldo. *A Sand County Almanac.* New York: Oxford University Press, 1949.

Lepage, Denis. "Townsend's Bunting in Ontario?" *Birding,* Vol. 46 (July/August 2014), 30–32.

Line, Les. "Silence of the Songbirds," *National Geographic,* Vol. 183 (June 1993), 68–91.

Lopez, Barry. *Arctic Dreams.* New York: Charles Scribner's Sons, 1986.

Luoma, Jon R. "Forests Are Dying, But Is Acid Rain Really to Blame?" *Audubon,* Vol. 89 (March 1987), 36–51.

Magnusson, Magnus, and Hermann Piilsson, trans. *The Vinland Sagas.* London: Penguin Books, 1965.

Marchand, Peter J. *Life in the Cold.* Hanover, NH: University Press of New England, 1987.

Matthiessen, Peter. *Wildlife in America.* Rev. ed. New York: Viking Press, 1987.

McMannamay, Rachel H., Lynn M. Resler, James B. Campbell and Ryan A. McManamay. "Assessing the Impacts of Balsam Woolly Adelgid (*Adelges piceae* Ratz.) and Anthropogenic Disturbance on the Stand Structure and Mortality of Fraser Fir (*Abies fraseri* [Pursh] Poir.) in the Black Mountains, North Carolina." *Castanea,* Vol. 76 (March 2011), 1–19.

McRaven, Charles. "The Survivor," *Country Journal,* Vol. 27 (May–June 1990), 34–39.

Meades, Susan J. "The Barrens: Heathlands of Newfoundland," Wildflower (Winter 1993), 32-35.

Merritt, Joseph F. Guide to the Mammals of Pennsylvania. Pittsburgh: University of Pittsburgh Press/Carnegie Museum of Natural History, 1987.

Mitchell, John G. "Lord of the Eastern Forests," Audubon, Vol. 90 (March 1988), 58-61.

———."Love and War in the Big Woods," Wilderness, Vol. 55 (Spring 1992), 1022-31.

Mohlenbrock, Robert H. "Blackies Hollow, Virginia" (This Land), Natural History, Vol. 97 July 1987), 71-72.

———. "Gaspesie Park, Quebec" (This Land), Natural History, Vol. 101 (June 1992), 22-25.

———. "Flagpole Knob, Virginia" (This Land), Natural History, Vol. 102 (January 1993), 22-24.

———. "Forillon National Park, Quebec" (This Land), Natural History, Vol. 102 (May 1993), 74-76.

Monastersky, R. "Acid Precipitation Drops in United States," *Science News*, Vol. 144 (July 10, 1993), 22.

Moore, Harry L. *A Roadside Guide to the Geology of the Great Smoky Mountains National Park.* Knoxville: University of Tennessee Press, 1988.

Morison, Samuel Eliot. *The European Discovery of America: The Northern Voyages.* New York: Oxford University Press, 1971.

Morse, Douglass H. *American Warblers.* Cambridge, MA: Harvard University Press, 1989.

Morton, Eugene S. "What Do We Know About the Future of Migrant Songbirds?" in *Ecology and Conservation of Neotropical Migrant Landbirds,* John M. Hagan III and David W. Johnston, eds. Washington, DC: Smithsonian Institution Press, 1992.

Morton, Eugene S., and R. Greenberg. "Outlook for Migratory Songbirds: Future Shock for Birders," *American Birds,* Vol. 43 (Spring 1989), 178–83.

Mountfort, Guy. *Rare Birds of the World.* Lexington, MA: The Stephen Greene Press, 1988.

Napier, Shelter. *Great Smoky Mountains.* Washington, DC: National Park Service/ Handbook 112, 1981.

Nowak, Ronald M. *Walker's Mammals of the World.* 5th ed. Baltimore: Johns Hopkins University Press, 1991.

———."The Red Wolf Is Not a Hybrid," *Conservation Biology,* Vol. 6 (December 1992), 593–95.

Nuss, Donald L. "Engineered Hypovirulence," *The Bark,* newsletter of American Chestnut Foundation (May–June 1993), 1–4.

Ogburn, Charlton. *The Southern Appalachians: A Wilderness Quest.* New York: William Morrow, 1975.

Opler, Paul A. *Eastern Butteiflies.* Boston: Houghton Mifflin Co., 1992.

Osborn, Stephen G., Avner Vengosh, Nathaniel R. Warner and Robert B. Jackson. "Methane Contamination of Drinking Water Accompanying Gas-well Drilling and Hydraulic Fracturing," *Proceedings of the National Academies of Science,* Vol. 108 (2011), 8172–8176.

Pauley, Thomas K. *Cheat Mountain Salamander* (Plethodon nettingi) *Recovery Plan: Technical/Agency Draft.* Newton Corner, MA: United States Fish & Wildlife Service, 1991.

Peattie, Donald Culross. *A Natural History of Trees of Eastern and Central North America.* 1948: Boston: Houghton Mifflin Co., 1991.

Peters, Harold S., and Thomas D. Burleigh. *The Birds of Newfoundland.* Boston: Houghton Mifflin Co., 1951.

Pielou, E. C. After the Ice Age. Chicago: University of Chicago Press, 1991.

Porneluzi, Paul, J. Bednarz, L. Goodrich, N. Zawada, and J. Hoover. "Reproductive Performance of Territorial Ovenbirds Occupying Forest Fragments and a Continuous Forest in Pennsylvania," Conservation Biology, Vol. 7 (September 1993), 618–22.

Pringle, Larry. "Pennsylvania's Mystery Rabbit," Pennsylvania Game News, Vol. 34 (July 1963), 15–17.

Proctor, Noble S., and Patrick J. Lynch. Manual of Ornithology. New Haven, CT: Yale University Press, 1993.

Pyle, Robert Michael. Handbook for Butterfly Watchers. Boston: Houghton Mifflin Co., 1992.

Raloff, Janet. "Acid Rain's Most Visible Symptom," Science News, Vol. 137 (March 3, 1990).

———."Mercurial Risks from Acid's Reign," Science News, Vol. 139 (March 9, 1991), 152–56.

Raymo, Chet, and Maureen Raymo. Written in Stone. Chester, CT: Globe Pequot Press, 1989.

Redfern, Ron. The Making of a Continent. New York: Times Books, 1983.

Rennie, John. "Howls of Dismay," Scientific American (October 1991), 19–20.

Revkin, Andrew C. "Pressure Builds for Swift U.S. Action Against Spreading Salamander Threat." New York Times, May 15, 2015, http://dotearth.blogs.nytimes.com/2015/05/15/pressure-builds-for-swift-u-s-action-against-spreading-salamander-threat/.

Robbins, Chandler S., Danny Bystrak, and Paul H. Geissler. The Breeding Bird Survey: Its First 15 Years, 1965–1979. Washington, D.C.: United States Fish & Wildlife Service/Resource Publication 157, 1986.

Robin, G., ed. Biogeography and Ecology of the Island of Newfoundland. The Hague: Dr. W. Junk Publishers, 1983.

Rocks Adrift: The Geology of Gros Morne National Park. Canadian Ministry of the Environment, 1990.

Roe, Charles E. A Directory to North Carolina's Natural Areas. Raleigh: North Carolina Natural Heritage Foundation, 1987.

Roosevelt, Theodore. "Small Country Neighbors." Scribners Magazine, Vol. 42 (October 1907), 385–395.

Rue, Leonard L. Deer of North America. New York: Crown, 1979.

———. Furbearing Animals of North America. New York: Crown.

Sargent, Charles Sprague. Manual of the Trees of North America, Vol. I. New York: Dover Publications, 1965.

Schwarzkopf, S. Kent. A History of Mt. Mitchell and the Black Mountains. Raleigh: North Carolina Division of Archives and History, 1985.

Scott, Peter. "Origins of Newfoundland's Flora," Wildflower (Winter 1993), 16–18.

Severin, Timothy. "The Voyage of 'Brendan,'" National Geographic, Vol. 152 (December 1977), 770–97.

Smith, Ned. "Meadowcroft: Hunting Camp of the Ancients," *Pennsylvania Game News*, Vol. 47 (June 1976).

Smith, Richard. *Wild Plants of America.* New York: John Wiley & Sons, 1989.

Stegmann, Eileen C. "The Zebra Mussel—New York's Carpetbagger," New York *Conservationist*, Vol. 47 (July/August 1992), 37–41.

Steiner, Linda. "Building Better Bogs," *Pennsylvania Wildlife*, Vol. 11 (January/February 1990), 16–19.

Stewart, James B. "The Markets Dig a Grave for Big Coal," New York Times, August 7, 2015, A1.

Stolzenberg, William. "The Mussels' Message," *Nature Conservancy* (November/December 1992), 16–23.

Storey, Janet M., and Kenneth B. Storey. "Out Cold," *Natural History*, Vol. 101 (January 1992), 22–25.

Stupka, Arthur. Notes on the Birds of Great Smoky Mountains National Park. Knoxville: University of Tennessee Press, 1963.

Such, Peter. *Vanished Peoples: The Archaic, Dorset and Beothuck People of Newfoundland.* Toronto: NC Press Ltd., 1978.

Swanton, John R. *Final Report of the United States DeSoto Expedition Commission.* 1939. Reprint, Washington, D.C.: Smithsonian Institution Press, 1985.

Tanner, Ogden. *New England Wilds.* New York: Time-Life Books, 1974.

Terres, John K. *The Audubon Society Encyclopedia of North American Birds.* New York: Alfred A. Knopf, 1980.

Thoreau, H. D. *Faith in a Seed,* quoted in "The Forest in the Seeds," Barbara Kingsolver, *Natural History*, Vol. 102 (October 1993), 36–39.

Todd, W. E. Clyde. *Birds of Western Pennsylvania.* Pittsburgh: University of Pittsburgh Press, 1940.

Tome, Philip. *Pioneer Life, Or, Thirty Years a Hunter.* 1854. Reprint, Baltimore: Gateway Press, 1989.

Tramer, Elliot J. "Global Warming: An Imminent Threat to Birds," *Living Bird*, Vol. 11 (Spring 1992), 8–12.

"Tree Defenses Aiding Gypsy Moth Spread," *Pennsylvania Woodland News*, Pennsylvania State University Forest Resource Extension, Vol. 6 (January/February 1992), 1.

Tuck, James A. *Newfoundland and Labrador Prehistory.* Ottowa: National Museum of Man, 1976.

Tullar, Ben, Jr. "The Eastern Coyote: Always a New York Native," *New York Conservationist*, Vol. 46 (January/February 1992), 35–39.

Turner, Frederick. "In the Highlands," *Wilderness*, Vol. 53 (Fall 1990), 26–40, 70–76.

Tyning, Thomas A. *A Guide to Amphibians and Reptiles.* Boston: Little, Brown, 1990.

Udall, James R. "Launching the Natural Ark," *Sierra* (September/October 1991), 79–89.

Van Diver, Bradford B. *Roadside Geology of Pennsylvania.* Missoula, MT: Mountain Press Publishing, 1981.

Walker, Tim. "Dreissena Disaster," *Science News*, Vol. 139 (May 4, 1991), 282–84.

Wallace, Paul A. W. *The Indians of Pennsylvania*. Harrisburg, PA: Pennsylvania Historical and Museum Commission, 1968.

Weidensaul, Scott. "Return of the Elk," *Smithsonian* (December 1999), 82–94.

Welty, Joel C., and Luis Baptista. *The Life of Birds*. 4th ed. Philadelphia: W. B. Saunders, 1988.

Wharton, Donald. "Return of the Black Cat," *New York Conservationist* (March/April 1992), 37–39.

Wilderman, Dr. Candie C. *Acid Precipitation*. Harrisburg, PA: Pennsylvania Fish Commission, 1989.

Wilcove, David S. "Where Have All the Songbirds Gone?" *Living Bird Quarterly*, Vol. 4 (Spring 1985), 20–23.

———."Empty Skies," *Nature Conservancy*, Vol. 40 (January 1990), 4–13.

———. "Crossbills and Clearcuts," *Living Bird*, Vol. 12 (Autumn 1993), 34–35.

Williams, Harold. *Tectonic-Lithofacies Map of the Appalachian Orogeny*. St. John's: Memorial University of Newfoundland, 1978.

Willis, Edwin O. "Columbus and the River of Birds," *Natural History*, Vol. 101 (October 1992), 22–29.

Wilson, Edward O. *The Diversity of Life*. Cambridge, MA: Belknap Press/Harvard University Press, 1992.

Wilson, Samuel M. "The Vikings and the Eskimo," *Natural History*, Vol. 101 (February 1992), 18–21.

Wuerthner, George. *Southern Appalachian Country*. Helena, MT: American Geographic Publishing, 1990.

Young, Matthew A., Ken Blankenship, Marilyn Westphal and Steve Holzman. "Status and Distribution of Type 1 Red Crossbill (*Loxia curvirostra*): An Appalachian Call Type?" *North American Birds*, Vol. 65 (2011), 554–561.

Young, Matthew A., David A. Fields and William A. Montevecchi. "New Evidence in Support of a Distinctive Red Crossbill (*Loxia curvirostra*) Type in Newfoundland," *North American Birds*, Vol. 66 (2012), 29–33.

Index

About the Author

Author and naturalist **Scott Weidensaul** has written more than two dozen books on natural history, including Pulitzer Prize finalist *Living on the Wind: Across the Hemisphere with Migratory Birds, The Ghost with Trembling Wings, Of a Feather: A Brief History of American Birding,* and *The First Frontier: The Forgotten History of Struggle, Savagery and Endurance in Early America.*

Weidensaul lectures widely on wildlife and environmental topics, and is an active field researcher, specializing in birds of prey and hummingbirds. He lives in the Appalachians of eastern Pennsylvania.